SMP interact
for GCSE

Book I1 part B

I1
PART B

PATHFINDER
EDITION

PUBLISHED BY THE PRESS SYNDICATE OF THE UNIVERSITY OF CAMBRIDGE
The Pitt Building, Trumpington Street, Cambridge, United Kingdom

CAMBRIDGE UNIVERSITY PRESS
The Edinburgh Building, Cambridge CB2 2RU, UK
40 West 20th Street, New York, NY 10011-4211, USA
10 Stamford Road, Oakleigh, VIC 3166, Australia
Ruiz de Alarcón 13, 28014 Madrid, Spain
Dock House, The Waterfront, Cape Town 8001, South Africa

http://www.cambridge.org

© The School Mathematics Project 2001
First published 2001

Printed in Italy by Rotolito Lombarda
Typeface Minion   *System* QuarkXPress®

A catalogue record for this book is available from the British Library

ISBN 0 521 01244 9   paperback

Typesetting and technical illustrations by The School Mathematics Project
Illustrations on pages 34 (*A* and *B*) and 139 by Dave Parker and
page 34 (*C* and *D*) by David Parkins
Photographs by Graham Portlock

Acknowledgements

The authors and publishers are grateful to the following Examination
Boards for permission to reproduce questions from past examination
papers:

AQA(NEAB)	Assessment and Qualifications Alliance
AQA(SEG)	Assessment and Qualifications Alliance
Edexcel	Edexcel Foundation
OCR	Oxford, Cambridge and RSA Examinations
WJEC	Welsh Joint Education Committee

Contents

17 Sequences

You should know how to find the value of simple expressions such as $10 - 2n$, $n^2 + 3$

You will

◆ form sequences of numbers such as odds, evens, squares and triangle numbers

◆ find and use rules to continue a variety of sequences

◆ find and use rules for the nth term of linear and simple non-linear sequences

◆ find a sequence from a context, find a rule for the nth term and explain how you found it

🄰 Sequences from shapes

These designs are made by arranging counters in squares.

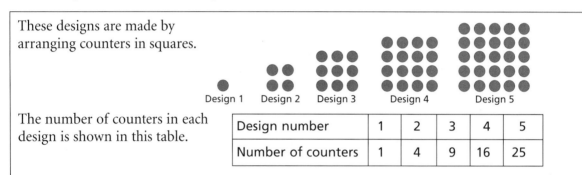

Design 1   Design 2   Design 3   Design 4   Design 5

The number of counters in each design is shown in this table.

Design number	1	2	3	4	5
Number of counters	1	4	9	16	25

The numbers in the sequence 1, 4, 9, 16, 25, … are called **square numbers**.

A1 (a) Find the 6th square number.

(b) What is the 20th square number?

A2 These designs are made by arranging counters in L-shapes.

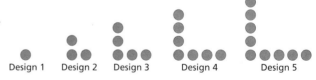

Design 1   Design 2   Design 3   Design 4   Design 5

(a) Copy and complete this table for these designs.

Design number	1	2	3	4	5
Number of counters	1				

(b) How many counters are in the 6th design?

(c) How many counters are needed to make the 15th design?
Explain how you worked out your answer.

(d) Which design uses 99 counters?

(e) Is it possible to make one of these designs with 40 counters? Explain your answer.

A3 These designs are made by arranging counters in triangles.

Triangle 1 Triangle 2 Triangle 3 Triangle 4 Triangle 5

(a) Draw the next two designs.

(b) Copy and complete this table.

Design number	1	2	3	4	5	6
Number of counters			6			

(c) The numbers of counters are called **triangle numbers**.
So, for example, the third triangle number is 6.

List the first ten triangle numbers.

A4 These lattice designs are all models of salt crystals.
They are cube-shaped.

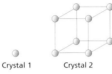 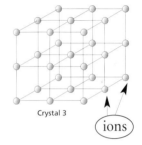

Crystal 1   Crystal 2   Crystal 3   ions

(a) Work out the number of ions in Crystal 3 without counting them all.

(b) Copy and complete this table.

Crystal number	1	2	3
Number of ions		8	

(c) Without drawing, work out how many ions will be in Crystal 4?

(d) The numbers of ions are called cubic numbers or cubes.
So, for example, the second cubic number is 8.
What is the 10th cubic number?

B *Following a rule*

In sequences of numbers, there is often a rule to go from one term to the next.
Some examples are

Sequence	**Rule**
8, 16, 24, 32, 40, …	Add 8 to the previous term
8, 16, 32, 64, 128, …	Multiply the previous term by 2
4, 5, 7, 10, 14, …	Add 1, then add 2, then add 3 and so on
4, 5, 9, 14, 23, …	Add the previous two terms together

Sequences where the rule is to **add or subtract the same amount** each time are **linear**.

B1 A sequence of numbers begins   5, 9, 13, …
The rule to continue this sequence is 'add 4 to the previous term'.
(a) What is the next term?   (b) What is the 8th term?
(c) Is the sequence linear?

B2 A sequence of numbers begins   5, 9, 17, …
The rule to continue this sequence is 'multiply the previous term by 2 and subtract 1'.
(a) What is the next term?   (b) What is the 6th term?
(c) Is the sequence linear?

B3 (a) Write down the next two numbers in this sequence   4, 7, 10, 13, 16, …
(b) Write down a rule to find the next two numbers.

B4 Write down the next two terms in the sequence   0, 2, 6, 14, …

B5 A sequence of numbers begins   3, 4, 6, 9, 13, 18, …
(a) Γ scribe a rule to go from one term to the next.
(b) U. ng your rule, what is the 7th term of this sequence?
(c) Is the sequence linear?

B6 For each of the following sequences
• describe a rule to go from one term to the next
• find the 8th term

(a) 1, 8, 15, 22, 29, …   (b) 35, 30, 25, 20, 15, …   (c) 1.5, 3, 6, 12, 24, …
(d) 800, 400, 200, 100, … (e) $\frac{1}{9}, \frac{1}{3}, 1, 3, 9, …$ (f) 1, 4, 13, 40, 121, …

B7 The first four terms of a sequence are   1, 4, 7, 10, …
For this sequence, explain how you can work out the value of the 30th term.

B8 Copy each sequence and fill in the missing numbers.
(a) 5, 7, 9, __ , 13, __ , 17, 19, …   (b) 1, 3, 7, 13, 21, 31, __ , 57, __ , …
(c) __ , 0.5, 1, 2, 4, 8, __ , 32, …   (d) 1, 1, 2, 3, 5, 8, 13, __ , 34, 55, __ , …
(e) 1, 5, 13, 29, __ , 125, __ , …

B9 Each sequence below is linear.
Copy each sequence and fill in the missing numbers.
(a) 4, 6, __ , __ , __ , 14, …   (b) 25, __ , 19, __ , 13, 10, …
(c) 1, __ , 11, __ , 21, __ , …   (d) __ , 10, __ , __ , 19, __ , …

*B10 Find the missing expression in each linear sequence below.
(a) x, $x + 4$, _____ , $x + 12$, …   (b) a, $a + b$, $a + 2b$, _____ , …
(c) $n - 5$, $n - 3$, $n - 1$, _____ , … (d) x, $x - y$, _____ , $x - 3y$, …

ℂ *The nth term*

We can work out any term of a sequence if we have an expression for its *n*th term.

Example

The *n*th term of a sequence is $2n + 5$. Find the first six terms.

The **1**st term is $2 \times \mathbf{1} + 5 = \mathbf{7}$
The **2**nd term is $2 \times \mathbf{2} + 5 = \mathbf{9}$
The **3**rd term is $2 \times \mathbf{3} + 5 = \mathbf{11}$ … and so on

We can show our results in a table.

Term numbers (n)	1	2	3	4	5	6	...
Terms of the sequence (2n + 5)	7	9	11	13	15	17	...

C1 Linear sequences can be found on this grid.
Two are shown on the diagram.

(a) Find seven more linear sequences
   that have four terms or more.

   Write down each sequence as an
   increasing sequence and find
   its next term.

(b) The expressions below give the *n*th terms
   of these sequences.

   Match each expression to its sequence.

44	34	24	14	4	3	6	9	12
40	30	5	20	10	11	5	8	1
44	37	30	23	16	9	2	7	3
4	11	23	21	22	12	1	6	9
1	7	26	20	28	9	8	5	0
3	31	10	15	34	30	12	4	8
36	6	11	13	40	0	1	3	2

| $3n$ | | $6n - 2$ | | $10n - 6$ | | $2n + 1$ | | $4n$ |

| | $3n + 1$ | | $n + 2$ | | $5n + 1$ | | $7n - 5$ | |

C2 An expression for the *n*th term of a sequence is $4n - 3$.
Work out the fourth and fifth terms of the sequence.

C3 The *n*th term of a sequence is $2n + 3$.

(a) Write down the first six terms of the sequence.

(b) Calculate the 100th term

C4 The *n*th terms of six different sequences are:

A $7n - 2$ B $10 - n$ C $\frac{1}{2}n - 3$ D $n^2 + 1$ E $\frac{60}{n}$ F 2^n

(a) Calculate the first five terms of each sequence.

(b) Calculate the 20th term of each sequence.

(c) Which of these sequences are linear?

Ⓓ *The nth term of a linear sequence*

Find the *n*th term of the linear sequence   3, 7, 11, 15, 19, …

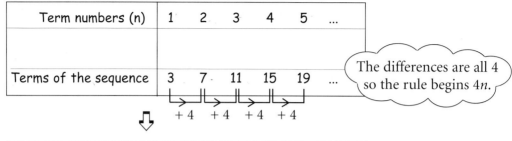

Term numbers (n)	1	2	3	4	5	…
Terms of the sequence	3	7	11	15	19	…

+4   +4   +4   +4

The differences are all 4 so the rule begins 4*n*.

Term numbers (n)	1	2	3	4	5	…
4n	4	8	12	16	20	
Terms of the sequence	3	7	11	15	19	…

− 1

To get from 4 to 3 you can subtract 1. This works for all terms.

So a rule for the *n*th term of the linear sequence   3, 7, 11, 15, 19, … is 4*n* − 1.

D1 For each of the following sequences

- find an expression for the *n*th term
- use your expression to work out the 50th term

(a) 4, 7, 10, 13, 16, …   (b) 1, 10, 19, 28, 37, …   (c) 2, 7, 12, 17, 22, …

(d) 4, 9, 14, 19, 24, …   (e) 3, 5, 7, 9, 11, …

D2 This diagram shows house numbers on North Street.

(a) What is the number of the 15th house ?

(b) Find an expression for the number of the *n*th house on North Street.

This diagram shows house numbers on South Street.

(c) What is the number of the 10th house?

(d) Find an expression for the number of the *n*th house on South Street.

(e) What are the house numbers of the 50th houses in each street?

D3 (a) Copy and complete this table for the linear sequence   40, 38, 36, 34, 32, … .

Term numbers (n)	1	2	3	4	5	...
$-2n$	-2	-4	?	?	?	
		?				
Terms of the sequence	40	38	36	34	32	...

The differences are all -2 so look at $-2n$.

(b) What is an expression for the nth term of the sequence?

(c) Calculate the 20th term in the sequence.

D4 For each of the following sequences, find an expression for the nth term.

(a) 30, 28, 26, 24, 22, …   (b) 40, 37, 34, 31, 28, …

(c) 33, 28, 23, 18, 13, …   (d) 60, 54, 48, 42, 36, …

E *Not just linear sequences*

Each of the expressions below gives the nth term of a sequence.

- Investigate these sequences.

A $\boxed{n^2}$ B $\boxed{n^2 + 4}$ C $\boxed{n^2 - 2}$ D $\boxed{2n^2}$ E $\boxed{2n^2 + 3}$ F $\boxed{5n^2}$

Find the nth term of the sequence   3, 6, 11, 18, 27, …

Term numbers (n)	1	2	3	4	5	...
Terms of the sequence	3	6	11	18	27	...

$+3 \quad +5 \quad +7 \quad +9$

The differences form a **linear** sequence 3, 5, 7, 9, … so the rule involves n^2.

Term numbers (n)	1	2	3	4	5	...
n^2	1	4	9	16	25	
	+2					
Terms of the sequence	3	6	11	18	27	...

To get from 1 to 3 you can add 2. This works for all terms.

So a rule for the nth term of the sequence   3, 6, 11, 18, 27 … is $n^2 + 2$.

E1 A sequence of numbers begins   2, 5, 10, 17, 26, 37, …

(a) What is the next term in the sequence?

(b) Explain how you can tell that the sequence is not linear.

(c) What is an expression for the nth term of this sequence?

E2 (a) A sequence of numbers begins   4, 7, 12, 19, 28, …

What is an expression for the nth term of this sequence?

(b) Calculate the 15th term.

E3 (a) A sequence of numbers begins   3, 12, 27, 48, 75, …

What is an expression for the nth term of this sequence?

(b) Calculate the 10th term.

E4 (a) A sequence of numbers begins   2, 8, 18, 32, 50, 72, …

What is an expression for the nth term of this sequence?

(b) What is the nth term of the sequence   3, 9, 19, 33, 51, 73, …

E5 For each of the following sequences, find an expression for the nth term.

(a) 0, 3, 8, 15, 31, …

(b) 11, 14, 19, 26, 35, …

(c) 4, 16, 36, 64, …

(d) 5, 17, 37, 65, …

𝔽 *Ways of seeing*

Some disco light units are designed as shown. They use red and yellow lights.

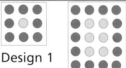

Design 1

Design 2

Design 3

The diagrams below show how some pupils counted the red lights in Design 5.

Ken

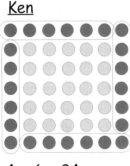

$4 \times 6 = 24$

Gill

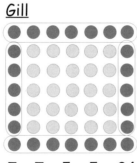

$7 + 7 + 5 + 5 = 24$

Raj

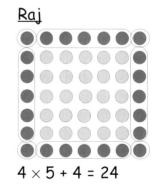

$4 \times 5 + 4 = 24$

They each found a rule for the number of red lights in Design n.

$r = (n + 2) + (n + 2) + n + n$

$r = 4n + 4$

$r = 4(n + 1)$

Who do you think found each rule?

F1 Here is another design for square light units.

Design 1

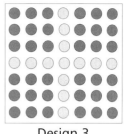

Design 2

Design 3

(a) Draw a diagram to show Design 4.

(b) For Design 4, what is the number of yellow lights?

(c) How many yellow lights would you need for Design 10?

(d) Find a rule for the number of yellow lights in Design n. Explain how you found your rule.

(e) How many yellow lights would you need for Design 50?

(f) Find a rule for the number of red lights in Design n.

F2 Here are some rectangular units.

Design 1

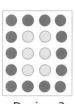

Design 2

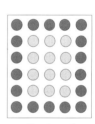

Design 3

(a) For Design 3, what is the number of red lights?

(b) How many red lights would you need for Design 8?

(c) Find a rule for the number of red lights in Design n. Explain how you found your rule.

(d) How many red lights would you need for Design 100?

(e) Find a rule for the number of yellow lights in Design n.

*F3 Here are some triangular units.

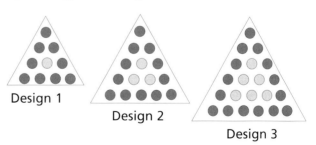

Design 1

Design 2

Design 3

(a) Find a rule for the number of red lights in Design n. Explain how you found your rule.

(b) How many red lights would you need for Design 100?

*G *Ways of seeing further*

TG

How many tins are in the 100th stack …
the *n*th stack?

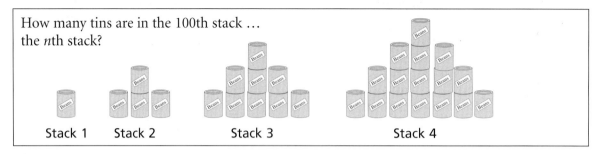

Stack 1   Stack 2   Stack 3   Stack 4

G1 For each set of patterns below:
- draw Pattern 5
- find a rule for the number of black tiles in the *n*th pattern
- find a rule for the number of white tiles in the *n*th pattern.

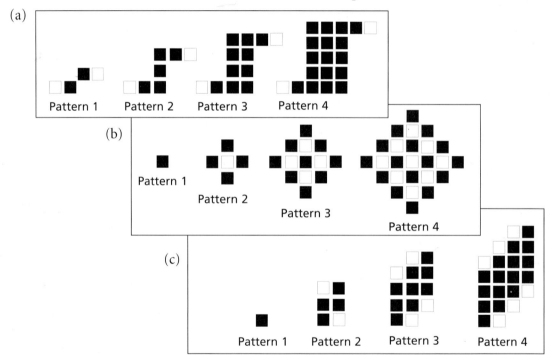

(a)

Pattern 1   Pattern 2   Pattern 3   Pattern 4

(b)

Pattern 1

Pattern 2

Pattern 3

Pattern 4

(c)

Pattern 1   Pattern 2   Pattern 3   Pattern 4

*H *Seeing even further*

TG

How many tins are in the 100th stack …
the *n*th stack?

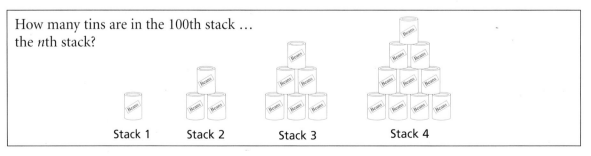

Stack 1   Stack 2   Stack 3   Stack 4

H1 The diagrams below show how to draw a 'mystic rose'.

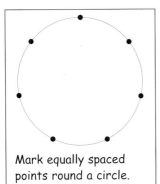

Mark equally spaced points round a circle.

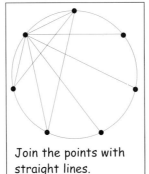

Join the points with straight lines.

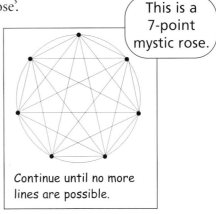

Continue until no more lines are possible.

This is a 7-point mystic rose.

(a) Draw three different mystic roses.
How many straight lines are in each of your designs?

(b) How many lines would be in a 20-point mystic rose?

(c) Find a rule for the number of lines in a n-point mystic rose.

Test yourself with these questions

T1 Write down the next two terms in the sequence 10, 11, 13, 16, 20, …

T2 (a) Find the eighth term of the sequence whose nth term is $4n - 1$.

(b) Find the nth term of the sequence whose first four terms are

2 8 14 20   OCR

T3 Each of these patterns uses black tiles.

(a) How many black tiles will in pattern 5?

(b) How many black tiles will be in pattern 10?

Pattern 1   Pattern 2   Pattern 3   Pattern 4

(c) What is an expression for the number of black tiles in pattern n?
Explain how you found your expression.

(d) How many black tiles will be in pattern 100?

T4 A sequence of numbers is shown below.

4 7 12 19 28 39

(a) Write down the next term in the sequence.

(b) Find an expression for the nth term of this sequence.

(c) Calculate the 30th term of the sequence.

AQA(SEG)1999

Percentage

For this work you need to know how to find equivalent fractions.

With and without a calculator, you will

◆ change between fractions, decimals and percentages

◆ find a percentage of a quantity

◆ write one quantity as a percentage of another

A Percentages, decimals and fractions

A scale can help you find equivalent decimals, fractions and percentages.
The scale is on sheet P48.

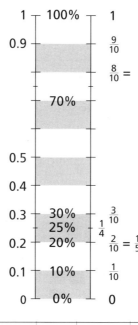

Some facts you need to know

$50\% = \frac{1}{2} = 0.5$

$25\% = \frac{1}{4} = 0.25$

$75\% = \frac{3}{4} = 0.75$

$1\% = \frac{1}{100} = 0.01$

$10\% = \frac{1}{10} = 0.1$

$20\% = \frac{1}{5} = 0.2$

$33\frac{1}{3}\% = \frac{1}{3} = 0.333\ldots = 0.\dot{3}$

Other facts you can work out

$5\% = \frac{5}{100} = \frac{1}{20}$

$40\% = \frac{40}{100} = \frac{4}{10} = 0.4$

$\frac{4}{5} = \frac{80}{100} = 80\%$

$\frac{7}{20} = \frac{35}{100} = 0.35$

$\frac{56}{80} = \frac{7}{10} = \frac{70}{100} = 70\%$

$0.41 = \frac{41}{100} = 41\%$

$0.04 = \frac{4}{100} = \frac{1}{25}$

$16\frac{1}{2}\% = \frac{16.5}{100} = 0.165$

$0.745 = \frac{745}{1000} = \frac{74.5}{100} = 74.5\%$

A1

A	C	D	E	G	H	I	M	N	O	P	R
$\frac{3}{4}$	5%	0.01	10%	$\frac{2}{5}$	80%	$33\frac{1}{3}\%$	0.5	25%	$\frac{1}{5}$	$\frac{3}{10}$	0.9

Use this code to find a letter for each fraction, decimal or percentage below.
Rearrange each set of letters to spell a city.

(a) $\frac{1}{2}$ $\frac{10}{100}$ $\frac{9}{10}$ 0.2

(b) $\frac{1}{100}$ 1% $\frac{1}{3}$ 0.75 90% 50%

(c) $\frac{75}{100}$ 0.4 $\frac{5}{100}$ $\frac{1}{20}$ $0.333\ldots$ $\frac{80}{100}$ $\frac{2}{10}$

(d) 30% $\frac{8}{10}$ 40% $\frac{25}{100}$ 20% $\frac{1}{10}$ $\frac{1}{4}$ 0.05 75% 0.1

A2 Try this without looking at the information on the previous page.

A	D	E	I	J	K	L	M	N	O	R	S	T	V	W
$\frac{1}{2}$	$\frac{4}{100}$	40%	0.75	45%	$\frac{1}{5}$	6%	0.01	0.8	$\frac{5}{100}$	$\frac{1}{3}$	$\frac{1}{4}$	60%	0.3	10%

Use this code to find a letter for each fraction, decimal or percentage below.
Rearrange each set of letters to spell a mountain.

(a) 25% $\frac{4}{10}$ $\frac{2}{5}$ 0.4 0.6 $0.\dot{3}$ 30%

(b) 80% 0.1 $\frac{8}{10}$ 4% 0.25 5% $\frac{1}{20}$

(c) $\frac{4}{5}$ 1% 75% 20% 50% $\frac{3}{4}$ 0.06 0.45 0.5 $33\frac{1}{3}$% 0.05

A3 Write these percentages as decimals

(a) 99%   (b) 3%   (c) 16%   (d) 7%   (e) 70%

(f) 30%   (g) 49.5%   (h) 12.6%   (i) 2.5%   (j) $12\frac{1}{2}$%

A4 Write these percentages as fractions, simplifying where possible.

(a) 60%   (b) 35%   (c) 48%   (d) 8%   (e) 33%

A5 Write these fractions as percentages.

(a) $\frac{7}{10}$ (b) $\frac{42}{50}$ (c) $\frac{7}{25}$ (d) $\frac{3}{5}$ (e) $\frac{9}{20}$

A6 Copy and complete $\dfrac{6}{40} = \dfrac{\blacksquare}{20} = \dfrac{\blacksquare}{100} = \blacksquare\,\%$

A7 Write these fractions as percentages.

(a) $\frac{18}{40}$ (b) $\frac{48}{80}$ (c) $\frac{27}{30}$ (d) $\frac{64}{200}$ (e) $\frac{9}{12}$

A8 Write $\frac{1}{8}$ as a percentage.

A9 Write these decimals as percentages.

(a) 0.55   (b) 0.07   (c) 0.8   (d) 0.375   (e) 0.015

A10 Find six matching pairs. Which card is left unmatched?

A $33\frac{1}{3}$%   **B** $\frac{3}{5}$ **C** 0.54   **D** $\frac{8}{25}$ **E** $\frac{1}{3}$ **F** 75%   **G** $\frac{27}{50}$

H 60%   **I** 0.8   **J** 8%   **K** 0.32   **L** $\frac{8}{100}$ **M** $\frac{6}{8}$

A11 Match these cards in groups of three.

A 0.65   **B** $\frac{3}{8}$ **C** 0.3   **D** 65%   **E** 0.375   **F** $\frac{12}{40}$

G $\frac{45}{50}$ **H** 30%   **I** 0.9   **J** $\frac{13}{20}$ **K** 90%   **L** $37\frac{1}{2}$%

A12 For each set of decimals, percentages and fractions, put them in order, smallest first.

(a) 0.5 $\frac{1}{10}$ 49% $\frac{1}{4}$ 20%   (b) $\frac{4}{5}$ 76% 0.9 0.08 5%

(c) $\frac{1}{3}$ $\frac{2}{5}$ 45% 33% 0.03   (d) $\frac{1}{5}$ 5% 0.4 4% 51%

Ⓑ *Percentage of a quantity*

100% is 84

50% is ?   25% is ?   10% is ?   5% is ?   15% is ?

75% is ?   60% is ?   45% is ?

B1 Pete has 32 apple trees.
25% of them are diseased.

How many of Pete's apple trees are diseased?

B2 In a class of 30 students, 10% have a weekend job.
How many students is this?

B3 75% of workers in a factory are full-time.
There are 120 workers. How many work full-time?

B4 A farmer has 60 sheep.
One year 40% of them have lambs.

How many have lambs?

B5 Sarah spent 10% of her £3.50 allowance on sweets. How much was this?

B6 Pete had £18 and spent 5% of his money. How much was this?

B7 Rasha saved 25% of her weekly allowance of £5. How much did she save?

B8 Work out

(a) 50% of £45   (b) 75% of £24   (c) 5% of 40p   (d) 25% of 60p

(e) 10% of £4.10   (f) 20% of £3   (g) 30% of £8   (h) 60% of £5

B9 Which is greater, and by how much: 10% of £20 or 15% of £12?

B10 Which is smaller, and by how much: 30% of £18 or 40% of £15?

B11 A cheesecake weighs 420 grams and 35% of the cheesecake is fat.
How many grams of fat are in the cheesecake?

B12 Pair up these amounts. Which amount is unmatched?

A 20% of £15   **B** 10% of £18   **C** 5% of £60   **D** 30% of £6

E 15% of £40   **F** 25% of £16   **G** 80% of £5

B13 Work out

(a) 15% of 60 grams   (b) 5% of 50kg   (c) 35% of 70 m

B14 (a) Work out 5% of £45.

(b) How could you use your answer to (a) to find 95% of £45?

B15 Work out   (a) 95% of £38   (b) 90% of £10.40

B16 Work out $33\frac{1}{3}$% of

(a) £12   (b) 21 kg   (c) 450 m   (d) 213 cm

VAT (Value Added Tax) is added to many things before you buy them. The rate of VAT is $17\frac{1}{2}$%.

Here is one method of working out the VAT for £60.

Find $17\frac{1}{2}$% of £60

10% of £60 = £6

5% of £60 = £3

$2\frac{1}{2}$% of £60 = £1.50

So $17\frac{1}{2}$% of £60 = £10.50

B17 Find the VAT at $17\frac{1}{2}$% on £40.

B18 Find $17\frac{1}{2}$% of

(a) £300   (b) £28   (c) £130   (d) £12.80

1% of £1

1% of £1 = 1p
13% of £1 = 13p

1% of £4 = 4p
23% of £4 = 23 × 4p = 92p

B19 Find

(a) 1% of £7   (b) 6% of £7   (c) 32% of £7   (d) 8% of £8   (e) 4% of £15

B20 (a) Find 1% of £24   (b) Find 99% of £24   (b) Find 2% of £5

(d) find 98% of £5   (e) Find 99% of £75   (f) Find 98% of £120

*B21 Are these statements true? Can you explain why?

3% of £8 = 8% of £3

5% of £13 = 13% of £5

When finding percentages without a calculator, it is often easiest to find 10% first.

Examples

Find 30% of £12

10% of £12 = £12 ÷ 10 = £1.20
So 30% of £12 = £1.20 × 3 = £3.60

Find 45% of 60

10% of 60 = 60 ÷ 10 = 6

40% of 60 = 6 × 4 = 24
5% of 60 = 6 ÷ 2 = 3

So 45% of 60 = 24 + 3 = 27

Ⓒ *One number as a percentage of another*

To find one number as a percentage of another, you can start with a fraction and try to make the denominator 100.

Examples

Out of a class of 20 students, 13 of them watch 'Eastenders'. What percentage watch it?

$$\frac{13}{20} = \frac{65}{100} = 65\%$$

× 5 (applied to both)

Out of a class of 30 students, 6 of them don't eat breakfast. What percentage don't eat breakfast.

$$\frac{6}{30} = \frac{1}{5} = \frac{20}{100} = 20\%$$

÷ 6   × 5

C1 Out of a total of 20 people, 18 of them said they felt happier in the summer than in the winter. What percentage is this?

C2 In a survey, 24 out of 40 people chose chicken tikka massala as their favourite meal. What percentage is this?

C3 Ibrar sat three tests. His results were:

Geography   35 out of 50
French   32 out of 40
Science   17 out of 25

(a) Find his percentage mark for each test.

(b) In which test did he get the highest percentage?

C4 Write the following as percentages and put them in order of size, smallest first.

A 36 out of 60   **B** 200 out of 250   **C** 18 out of 45

D 14 out of 56   **E** 9 out of 12   **F** 21 out of 150

C5 This bag contains some different sweets.

12 Mintos
18 Toffees
15 Munchos
12 Choccos
3 Humbugs

(a) What percentage of the sweets are

    (i) Mintos   (ii) Toffees

    (iii) Humbugs

(b) The Toffees and Choccos are wrapped in red paper.
What percentage of the sweets are wrapped in red?

(c) Freda doesn't like Munchos.
What percentage of the sweets does she like?

(d) Dean only likes Mintos and Munchos.
What percentage does he like?

$\mathbb{D}$ *A canoe club*

> # Come to the Carolina Canoe Club.
>
> *We offer a range of exciting activities every Saturday.*
>
> *Open to all ages.*
> *We have 240 members.*

D1 50% of the members are juniors.
How many members is this?

D2 60 members are aged 16 to 21.
What percentage are in this age group?

D3 10% of the members go canoeing every Monday evening.
How many is this?

D4 36 of the members have a life-saving certificate.
What percentage is this?

D5 55% of the club members are male.
How many males are in the canoe club?

D6 80% attend the summer barbecue.
How many come to the barbecue?

D7 One fifth of the members can do an 'Eskimo roll'.

(a) What percentage is this?

(b) How many can do an 'Eskimo roll'?

D8 84 took part on the last marathon race.
What percentage took part in this race?

D9 Two thirds attend the Annual General Meeting of the club.
How many of the members go to the meeting?

D10 Three fifths have their own canoes.

(a) What percentage is this?

(b) How many have their own canoes?

E *Percentage of a quantity (with a calculator)*

To find a percentage of a quantity,
you can change the percentage
into a decimal and then multiply.

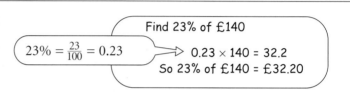

$23\% = \frac{23}{100} = 0.23$

Find 23% of £140

$0.23 \times 140 = 32.2$
So 23% of £140 = £32.20

E1 35% of Kim's bar of chocolate is fat.
The bar weighs 120 grams.
How much fat is in this bar of chocolate?

E2 85% of the 9700 households in a village own a car.
To the nearest hundred, how many households own a car?

E3 In a town of 15400 people, 32% own a bicycle but only 7% regularly cycle.
To the nearest hundred,

(a) How many people own a bicycle in this town?

(b) How many people regularly cycle?

E4 Work out

(a) 34% of 86   (b) 62% of 140   (c) 57% of 230

(d) 8% of 4500   (e) 12% of 32.5   (f) 90% of 3725

E5 82% of UK teenagers are worried about exams.
52% of UK teenagers worry about skin problems.

Out of a year group of 180, how many would you expect to

(a) worry about exams   (b) worry about skin problems

E6 In a town with a population of 56 000, it is estimated that 19% live on their own.
About how many people in this town live on their own?

E7 In many countries cattle or donkeys are used in harness for pulling ploughs or carts. This table shows the number of cattle and donkeys and the percentage used for work in harness, in various African countries.

Country	Cattle		Donkeys	
	Total number	% used in harness	Total number	% used in harness
Angola	3 100 000	10%	5 000	100%
Botswana	2 616 000	14%	152 000	92%
Mali	5 000 000	5%	550 000	27%
Senegal	2 740 000	5%	310 000	50%

(a) How many cattle work in harness in Mali?

(b) Which country uses the largest number of cattle in harness? Show how you decided on your answer.

(c) Which country uses the largest number of donkeys?

(d) Which country uses more donkeys than cattle?

E8 In a town of 15000 only 1.4% of the population cycle to work.

(a) Write 1.4% as a decimal.

(b) How many people in this town cycle to work?

E9 Work out

(a) 17.6% of £850   (b) 60.8% of 600 kg   (c) $3\frac{1}{2}$% of £5000

E10 Amanda has to pay a deposit of 2.25% on a car costing £13500. How much deposit does she have to pay?

F   *One number as a percentage of another (with a calculator)*

Example

Write the value as a fraction and change it to a decimal by dividing.

Multiply by 100 to change the decimal to a percentage.

Round your answer to one decimal place if necessary.

Rose earns £320 each week. One week, she spent £65 on food. What percentage of her earnings went on food?

$\frac{65}{320} = 65 \div 320 = 0.203125$

$0.203125 \times 100 = 20.3125$

So 20.3% of Rose's earnings went on food.

F1 In a school 115 of the 250 pupils in Year 10 own a bicycle. What percentage is this?

F2 The table shows how 1560 students travel to their school in the morning.

Copy and complete the table to show the percentage of pupils in each category.

Method of transport	Number of students	% of students
Bus	313	20.1%
Car	216	
Bicycle	377	
Walk	654	
Total	1560	

F3 In Kenya there are 420 000 motor vehicles and 32000 of these are motorbikes. What percentage of the motor vehicles in Kenya are motorbikes?

F4 In 1996, there were the following motor vehicles in the UK.

- 21 172 000 passenger vehicles
- 3 011 000 commercial vehicles
- 609 000 motorbikes

(a) How many motor vehicles were there in total in the UK in 1996?

(b) What percentage were passenger vehicles?

F5 This table shows the estimated number of donkeys in North Africa in 1996.

North African Country	Number of donkeys (thousands)
Algeria	230
Egypt	1690
Libya	55
Morocco	880
Tunisia	230
Total	3085

(a) How many donkeys are there in Egypt?

(b) What percentage of North African donkeys are found in Egypt?

(c) What percentage are found in Tunisia?

(d) Is it true that about 18% of North African donkeys are found in Libya?

ⓖ *Mixed questions*

G1 This pie chart shows information about how often adults used
a motor vehicle in Great Britain in 1995.

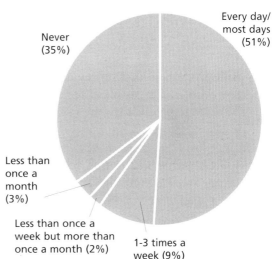

Driving a motor vehicle in Great Britain: 1995

Every day/most days (51%)

Never (35%)

Less than once a month (3%)

Less than once a week but more than once a month (2%)

1-3 times a week (9%)

(a) What percentage of people
drove at least once in 1995?

(b) A town had a population
of 17600 adults in 1995.

Find an estimate for

   (i) the number of these adults
that drove most days in 1995

  (ii) the number that drove
less than once a month

 (iii) the number that drove
once a week or more

G2 This table shows data on the use of the Channel Tunnel in 1996.

(a) How many UK residents
used the Channel Tunnel
in 1996?

(b) What percentage of UK
residents using the
tunnel were going
on holiday?

(c) What percentage of
people not resident
in the UK used the tunnel
for a business trip?

Use of the Channel Tunnel in 1996		
Purpose	UK residents (thousands)	Non-residents (thousands)
Holiday	1910	1458
Business	666	481
Visiting friends and relatives	299	498
Other	583	271
Total	3458	2708

(d) What percentage of all users of the tunnel were visiting friends and relatives?

G3 France has a total road network of 805070 kilometres.
Motorways make up a total of 6570 km.
90.0% of the total network is paved.

(a) What percentage of French roads are motorways?

(b) How many kilometres of French roads are unpaved?

Test yourself with these questions

T1 In a school council election, a total of 360 pupils voted.

90 pupils voted for Tanya Filton. What percentage voted for Tanya?

T2 A group of 40 women applied to go on an art trip.
Only 80% of all the women who applied went on the trip.

How many women went on the trip?

T3 A pupil carries out a survey in her town centre.
She chooses 200 people at random.
70 of these people say they use local swimming pool.

What percentage is 70 out of 200?

T4 400 members of a Sports Club are
asked, 'What is your favourite sport?'
The pie chart shows the results.

(a) What percentage chose tennis?

(b) How many members chose volleyball?

(c) Calculate the size of the angle for football.
Show all your working.

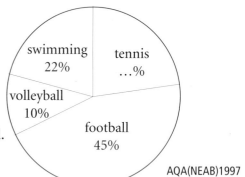

AQA(NEAB)1997

T5 There are 3270 students at a university.
At this university 1962 students have a part-time job.

What percentage of the students have a part-time job?

T6 In a survey it was found that 165 out of 440 people had been to France on holiday that year. What percentage of the people took a holiday in France?

T7 At Roughwood school there are 186 students in Year 11.
101 are boys. What percentage of the students are boys?

T8 One year, a garage sells 560 cars.
$62\frac{1}{2}$% of these cars were a shade of red.

How many of these cars were red?

T9 A carriage in a train has 60 seats. Only 1 of these is reserved for disabled passengers.
What percentage of seats are reserved for disabled passengers?

T10 In 1995 police carried out a total of 702.7 thousand breath tests.
13% of them were positive. How many was this, correct to the nearest thousand?

19 Graphs

Before you start this work, you should know how to

◆ work out a table of values for a straight-line graph

◆ plot a graph from a table
◆ read values from a graph

◆ form a formula from a practical situation
◆ interpret a graph of a practical situation

You will learn how to

◆ substitute into and interpret quadratic functions

◆ draw the graphs of quadratic equations and use them to solve simple problems

◆ use quadratic graphs and straight lines to solve related equations

A Straight lines review

$y = 2x - 1$

x	$^-2$	0	2
$2x - 1$	$^-5$	$^-1$	3

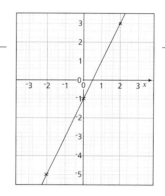

A1 A **linear** equation is one whose graph is a straight line.
Which of these are linear equations?

A $2y + 3x = 1$ B $y = 3x^2 - 1$ C $x + y = 1$ D $y = x^2 - 1$ E $y = 2x - 1$ F $y = x^2 + 5$

A2 (a) Copy and complete this table for $y = 2x + 3$.

(b) On graph paper, draw axes with x from $^-3$ to 3
and y from $^-4$ to 10.
Draw and label the graph of $y = 2x + 3$.

x	$^-2$	0	2
$2x + 3$			

A3 (a) Copy and complete this table for $y = 6 - 2x$.

(b) On graph paper, draw axes with x from $^-1$ to 5
and y from $^-4$ to 8.
Draw and label the graph of $y = 6 - 2x$.

x	0	2	4
$6 - 2x$			

A4 This question is on sheet P50.

A5 (a) Copy and complete this table of values for $y = \frac{1}{2}x - 2$.

x	-2	0	2
$\frac{1}{2}x - 2$	-3		

(b) On graph paper, draw axes with x from ⁻2 to 5 and y from ⁻4 to 4. Draw and label the graph of $y = \frac{1}{2}x - 2$.

A6 (a) Copy and complete this table of values for $y = 3$.

x	-3	-2	⁻1	0	1	2	3
y	3	3	3				

(b) Draw and label the graph of $y = 3$. Choose your own values on the axes.

A7 The diagram shows the graphs of four equations,

$$x = 3, \quad x = ^-1, \quad y = 3 \text{ and } y = ^-1.$$

Which line goes with which equation?

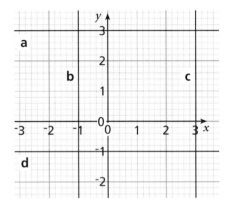

A8

The diagram shows four straight lines.

Write down the equation of each line.

A9 Choosing your own scales on the axes, draw and label the graphs of

(a) $y = 4$ (b) $y = ^-4$

(c) $x = 4$ (d) $x = ^-4$

To draw the graph of $x + 2y = 12$ you need to find points that fit the equation.

- Can you spot any values of x and y that fit?
- When x is 4, what value of y fits the equation? (You need to find y so that $4 + 2y = 12$.)
- Check that when $x = 0$, $y = 6$.
- Copy and complete this table.
- On graph paper draw axes with both x and y going from 0 to 12.
- Plot your points and draw the line $x + 2y = 12$

x	0	2	4	6	8	10	12
y	6						

A10 (a) To draw the line $x + y = 5$, we need to find some points that fit the equation $x + y = 5$.
When x is 3 the equation becomes $3 + y = 5$.
What is the value of y when $x = 3$?

(b) Work out the value of y when x is 0.

(c) What is the value of $\boldsymbol{x}$ when $\boldsymbol{y} = 0$?

(d) Copy and complete this table for $x + y = 5$.

(e) On graph paper, draw axes with x from $^-2$ to 7 and suitable values of y.
Draw the graph of $x + y = 5$.

(f) From your graph, what is the value of y when $x = ^-0.8$?

x	0	3	
y			0

A11 (a) Copy and complete this table for $3x + y = 6$.

(b) On graph paper, draw axes with x from $^-1$ to 3 and suitable values of y.
On your axes draw the graph of $3x + y = 6$.

x	0	1	
y			0

A12 (a) Copy and complete this table for $3x + 2y = 12$.

(b) Draw axes with x from $^-1$ to 5 and suitable values of y.
On your axes draw the graph of $3x + 2y = 12$.

(c) From your graph, what is the value of y when $x = 2.5$?
(Give your answer to one decimal place.)

x	0	2	
y			0

A13

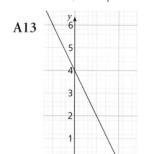

For the straight-line graph on the left

(a) Check that when $x = 0$, $y = 4$.

(b) Copy and complete this table of values for the line.

(c) Which of the equations below is the equation of the straight line?

x	y
0	4
1	
2	

    P $y = x + 1$ Q $y = 2x - 1$ R $x + y = 4$
    S $2x + y = 4$ T $x + y = 2$ U $x + 2y = 4$

Ⓑ *Problem review*

B1 You can put an advertisement into the *Evening news*.
The paper charges £25 to typeset the advert,
and then £4 for each centimetre of depth.

(a) What would the total cost be for this advert?

(b) How much would an advert 20 cm deep cost?

(c) Copy and complete this table for adverts.

Depth (*d*)	5	10	20
Cost in £ (*c*)			

(d) On graph paper, draw axes with *d* going across
from 0 to 20 and with *c* going up from 0 to 110.
Plot the points from your table and join them.

(e) Sue spends £75 on an advert.
What is the depth of her advert in cm?

B2 *Mendip Mushrooms* supply mushroom compost to gardeners.
They charge £35 delivery and then £6 per tonne of compost.

(a) How much would 4 tonnes of compost cost delivered to your door?

(b) How much would 1 tonne cost?

(c) Copy and complete this table for deliveries
of *Mendip Mushroom's* compost.

Weight (*w*)	1	4	6
Cost in £ (*c*)			

(d) Draw axes with *w* going across from 0 to 8 and *c* going up from 0 to 90.
Plot the points from your table and join them.

(e) Dave asks for £55 worth of compost to be delivered.
How much will he get, to the nearest $\frac{1}{10}$th of a tonne?

(f) Find a formula that connects *w* (the weight delivered in tonnes)
and *c* (the cost in £).

B3 *Fuming Fertilisers* also supply compost.
The formula they use for working out the cost of a delivery is $c = 10 + 10w$.
c is the cost in £, *w* is the weight in tonnes.

(a) Copy and complete the table on the
right for *Fuming Fertilisers'* prices.

Weight (*w*)	1	4	7
Cost in £ (*c*)			

(b) On the same axes you used for
question B2, draw the graph of $c = 10 + 10w$.

(c) What do *Fuming Fertilisers* charge for a delivery of 5.5 tonnes of compost?

(d) Use the graph to say which company would be cheaper for 10 tonnes of compost.
Explain your answer carefully.

ℂ *Quadratic graphs*

$y = x^2$

x	-2	-1.5	-1	-0.5	0	0.5	1	1.5	2
x^2	4								

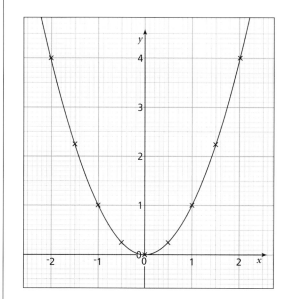
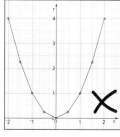
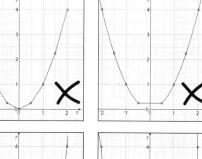
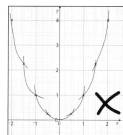
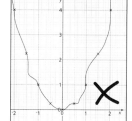

C1 (a) Copy and complete this table of values for $y = x^2 - 2$.

x	-2	-1	0	1	2
x^2	4	1			
$x^2 - 2$	2	-1			

    (b) On graph paper draw a pair of axes with x from -2 to 2, and y from -2 to 2. Draw the graph of $y = x^2 - 2$.

    (c) (i) At what values of x does the graph meet the x-axis? (Give your answers to 1 d.p.)

        (ii) Write down the two solutions to the equation $x^2 - 2 = 0$.

    (d) What value of x makes y smallest (a **minimum**)?

C2 (a) Copy and complete this table for $y = x^2 + x$.

x	-2	-1	0	1	2
x^2	4				
$x^2 + x$	2				

    (b) On graph paper draw a pair of axes. Draw the graph of $y = x^2 + x$.

    (c) (i) For what values of x is $y = 1$? (Give your answers to 1 d.p.)

        (ii) Write down the two solutions to the equation $x^2 + x = 1$.

    (d) On your graph draw the line of symmetry of $y = x^2 + x$. Write down the equation of the line of symmetry.

    (e) What value of x makes y a minimum?

C3 (a) Copy and complete this table for $y = 6 - x^2$.

x	-3	-2	-1	0	1	2	3
x^2		4					9
$6 - x^2$		2					-3

(b) On graph paper draw a pair of axes with x from $^-3$ to 3 and y from $^-4$ to 6.
On your axes, draw the graph of $y = 6 - x^2$.
Label the graph with its equation.

(c) Use your graph to solve the equation $6 - x^2 = 3$.

(d) What value of x makes y a maximum?

C4 (a) Copy and complete the table below for $y = 2x^2 - 7$.

x	-3	-2	-1	0	1	2	3
x^2	9	4					
$2x^2$	18	8					
-7	-7	-7	-7	-7	-7	-7	-7
$y = 2x^2 - 7$	11	1					

(b) Draw axes with x from $^-3$ to 3, and y from $^-8$ to 12.
Draw the graph of $y = 2x^2 - 7$ on your axes.

(c) (i) What values of x make $2x^2 - 7$ equal to 4?

(ii) Write down the solutions to the equation $2x^2 - 7 = 4$.

(d) Use your graph to solve the equation $2x^2 - 7 = 0$.

C5 This question is on sheet P52   OCR

C6 (a) Copy and complete the following table of values for $y = x^2 + 2x - 4$.

x	-4	-3	-2	-1	0	1	2	3
x^2								
$2x$								
-4	-4	-4	-4	-4	-4	-4	-4	-4
$y = x^2 + 2x - 4$								

(b) On suitable axes, draw the graph of $y = x^2 + 2x - 4$.

(c) What is the minimum value of $x^2 + 2x - 4$?

(d) Use your graph to solve the equation $x^2 + 2x - 4 = 0$.

(e) Solve the equation $x^2 + 2x - 4 = 1$.

C7 Julia throws a stone from the top of a cliff, as shown.

The equation of the path of the stone is $y = 50 - \dfrac{x^2}{2}$
(x and y are measured in metres.)

(a) Copy and complete this table for $y = 50 - \dfrac{x^2}{2}$.

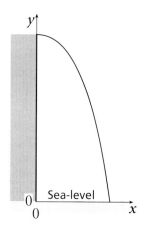

x	0	2	4	6	8	10	12
x^2	0	4	16	36			
$\frac{1}{2}x^2$	0	2	8	18			
$y = 50 - \frac{1}{2}x^2$	50	48	42	32			

(b) On graph paper, draw axes for x from 0 to 12 and for y from $^-30$ to 50. Plot the graph.

(c) What is the height of the cliff?

(d) How far from the bottom of the cliff does the stone hit the sea?

(e) From the graph, what is the value of y when x is 5? What does this tell you?

(f) Use the equation to work out the value of y when x is 16. Why is this information meaningless in this case?

D Fairground graphs

At Jeff's stall you can win a goldfish.
The goldfish are in different shaped bowls.

Jeff fills the bowls with water before he puts the goldfish in.
He uses a hose-pipe, from which water flows at a steady rate.

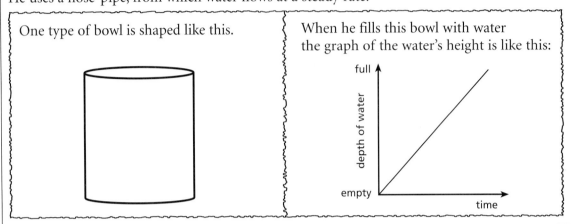

One type of bowl is shaped like this.

When he fills this bowl with water
the graph of the water's height is like this:

D1 Here are three different shaped bowls.

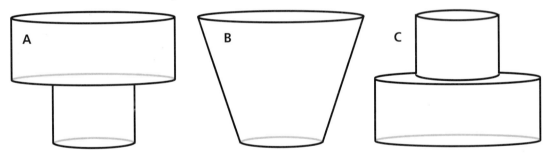

(a) Which description of filling the bowls with water goes with which bowl?

    P: The water level goes up fast at first and then suddenly goes up more slowly.

    Q: The water level goes up slowly at first, then changes to go up more quickly.

    R: The water level starts by going up quickly, but gets slower and slower.

(b) Which graph goes with which bowl?

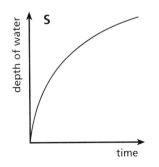 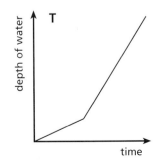 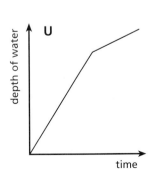

D2 Here are three more bowls.
For each one, sketch a graph showing how it fills up with water.

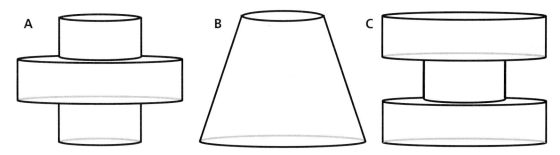

A   B   C

D3 The graph shows the number of people (customers and workers)
in the fairground one evening.

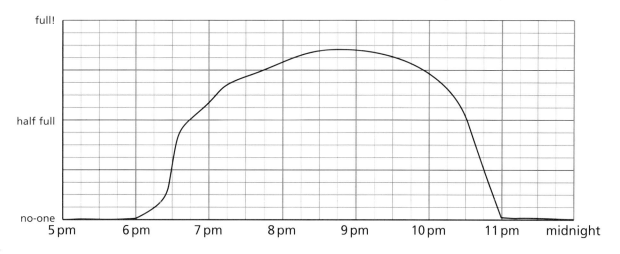

(a) At what time were there most people in the fairground?

(b) When do you think the fair opened?

(c) What time do you think the fair closed?

(d) At what time were most people arriving?

(e) Jeff is happy when the fair is more than half full.
For about how long was Jeff happy this evening?

D4 Look at the graphs at the bottom of the page.
Some of the graphs describe situations at the fair-ground.
Which graph describes which situation.

(Two graphs don't describe anything!)

Situations

A
The speed (*y*) of a
person against time
(*x*) as they come
down the
helter-skelter.

B
The height (*y*) of a
horse against time (*x*)
as the roundabout
goes round.

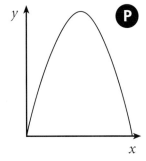

C
The height (*y*) of the
pinger on the 'try-
your-strength'
machine
against time
(*x*) when
someone
tries their
strength.

D
The total amount of
money taken (*y*)
against time (*x*) on
Jeff's stall.

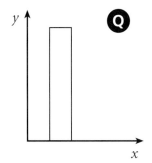

Graphs

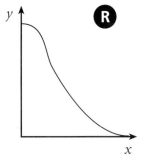

P

Q

R

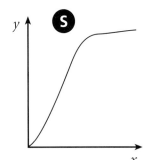

S

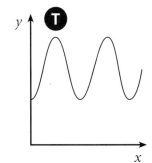

T

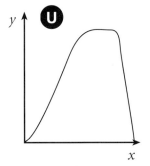

U

D5 Sketch a graph for each of these situations.
For each graph, write a short explanation of why it looks as you have drawn it.

(a) A dodgem car is going round the track at constant speed.
(Draw the dodgem car's speed on the y-axis against time on the x-axis.)

(b) A dart is thrown and then hits the dart board, 2 metres away.
(Draw the dart's speed (y) against time (x).)

(c) The big wheel goes round twice at constant speed.
(Draw the height of a person on the wheel (y) against time (x).)

Test yourself with these questions

T1 (a) Copy and complete this table for $y = 2x - 2$.

x	$^-1$	0	3
$2x - 2$			

(b) On graph paper, draw axes with x from $^-3$ to 3 and y from $^-4$ to 4.
Draw the graph of $y = 2x - 2$.

T2

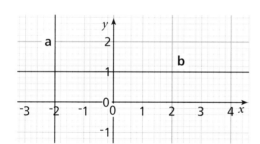

(a) Write down the equation of line a.

(b) What is the equation of line b?

T3 (a) Copy and complete this table for $y = x^2 - 3x$.

(b) On graph paper draw a pair of axes with x from $^-1$ to 4 and y from $^-3$ to 5.
Draw the graph of $y = x^2 - 3x$.

x	$^-1$	0	1	2	3	4
x^2	1	0		4	9	
^-3x	3	0		$^-6$	$^-9$	
$x^2 - 3x$	4	0		-2		

(c) Use your graph to solve the equations (i) $x^2 - 3x = 0$ (ii) $x^2 - 3x = 1$

(d) What value of x makes y a minimum?

T4 Draw sketch graphs for each of these situations.
For each graph, write a short explanation of why it looks as you have drawn it.

A A cyclist goes at a steady speed, then goes up a steep hill and stops at the top.
(Draw the cyclist's speed (y) against time (x).)

B The distance travelled on a motorway by a car going at constant speed.
(Draw the car's distance (y) against time (x).)

C The temperature of a saucepan of water that is heated up on a stove, and then left boiling for a short while.
(Draw the pan's temperature (y) against time (x).)

20 Using a calculator

You should know

◆ what brackets mean in a calculation

◆ how to round to a number of decimal places or significant figures

You will learn

◆ how to use a calculator for complex calculations

◆ how to work with squares, square roots and negative numbers

A Brackets and punctuation

| Four add two, multiplied by three | | Four, add two multiplied by three |

The position of the comma alters the calculation.

Brackets make the calculations clear:

| $(4 + 2) \times 3$ | | $4 + (2 \times 3)$ |

Scientific calculators do not need the brackets in the second calculation.
They automatically multiply or divide before they add or subtract.

| $4 + 2 \times 3 = 4 + 6 = 10$ |

A1 For each of the calculations below,

(i) predict what the result will be without using a calculator

(ii) then check with a calculator

(a) $7 + 5 \times 4$ (b) $5 \times (6 - 2)$ (c) $20 - (6 - 2)$ (d) $20 + 12 \times 4$

(e) $5 \times 6 + 3$ (f) $5 \times 3 + 7 \times 4$ (g) $5 \times (3 + 7) \times 4$ (h) $4 + 2 \times 8 + 3$

A2 Do these on a calculator.
Round each answer to 2 decimal places.

(a) $(4.82 + 2.94) \times 6.5$ (b) $4.82 + 2.94 \times 6.5$

(c) $3.74 \times 2.81 - 1.66$ (d) $12.65 - 2.91 \times 0.36$

(e) $8.64 + 2.37 \times 1.08 - 2.67$ (f) $0.85 \times (3.47 + 1.26) - 2.55$

(g) $4.22 \times 3.14 + 0.88 \times 2.57$

A3 Do these on a calculator.
Round each answer to 3 significant figures.

(a) $2.62 + 3.91 \times 4.5$ (b) $(1.82 + 4.94) \times 2.5$

(c) $40.4 \times (17.31 - 8.86)$ (d) $665 - 16.1 \times 13.2$

(e) $7.14 - 1.47 \times 1.13 + 4.61$ (f) $0.65 \times (3.52 - 1.46) + 3.58$

(g) $8.12 \times 0.64 - 0.92 \times 3.82$

B *Division*

In written calculations, a bar is often used for division.

written	on calculator	
$6 + \dfrac{24}{3}$	$6 + 24 \div 3$	Brackets not needed here. Calculator automatically does $24 \div 3$ first.
$\dfrac{6 + 24}{3}$	$(6 + 24) \div 3$	The division bar does a similar job to brackets.
$\dfrac{24}{7 + 3}$	$24 \div (7 + 3)$	
$\dfrac{6 + 24}{7 + 3}$	$(6 + 24) \div (7 + 3)$	

B1 Do these on your calculator.
All the answers should be whole numbers.

(a) $\dfrac{8.71 - 3.01}{1.9}$

(b) $\dfrac{130.9}{18.7} + 5$

(c) $(6.5 + 5.5) \times 1.5$

(d) $4.95 + 2.5 \times 2.02$

(e) $2.8 + \dfrac{2.88}{2.4}$

(f) $\dfrac{109.8}{4.4 + 1.7}$

(g) $\dfrac{17.38 + 2.22}{6.13 - 1.23}$

(h) $\dfrac{22.4}{0.76 + 0.64}$

(i) $\dfrac{14.08 - 6.93}{0.88 + 0.55}$

B2 Match each written calculation to a calculation on a calculator.

Written

A $5 + \dfrac{8}{2} + 3$

B $\dfrac{5}{8 + 2 + 3}$

C $\dfrac{5 + 8 + 2}{3}$

D $\dfrac{5 + 8}{2 + 3}$

E $5 + \dfrac{8 + 2}{3}$

F $\dfrac{5}{8 + 2} + 3$

G $\dfrac{5 + 8}{2} + 3$

Calculator

T $5 \div (8 + 2) + 3$

U $5 \div (8 + 2 + 3)$

V $5 + 8 \div 2 + 3$

W $(5 + 8) \div 2 + 3$

X $(5 + 8) \div (2 + 3)$

Y $(5 + 8 + 2) \div 3$

Z $5 + (8 + 2) \div 3$

B3 Calculate each of these, giving the result to 2 decimal places.

(a) $\dfrac{4.75 - 1.08}{2.03}$

(b) $0.68 + \dfrac{2.95}{1.07}$

(c) $\dfrac{4.86}{2.57 - 1.08}$

(d) $\dfrac{4.18 - 1.92}{7.15 - 3.28}$

(e) $\dfrac{115.4}{8.76 - 2.54}$

(f) $\dfrac{9.08 + 7.12}{6.48 - 3.25}$

C Using the memory

The memory on a calculator can sometimes be used instead of brackets.
Memory keys may be labelled 'M in', 'M out' or 'Store', 'Recall'.
For example, here are two ways to do $\dfrac{10.98}{4.4 + 1.7}$:

Using brackets   *10.98 ÷ (4.4 + 1.7)*

Using memory   *Do 4.4 + 1.7 first. Put the result in the memory.* $\boxed{4}\boxed{.}\boxed{4}\boxed{+}\boxed{1}\boxed{.}\boxed{7}\boxed{=}\boxed{\text{M in}}$

                     Then divide 10.98 by the number in the memory. $\boxed{1}\boxed{0}\boxed{.}\boxed{9}\boxed{8}\boxed{÷}\boxed{\text{M out}}\boxed{=}$

Some calculators have an 'ANS' key which recalls the last answer: $\boxed{4}\boxed{.}\boxed{4}\boxed{+}\boxed{1}\boxed{.}\boxed{7}\boxed{=}$

                                                 $\boxed{1}\boxed{0}\boxed{.}\boxed{9}\boxed{8}\boxed{÷}\boxed{\text{ANS}}\boxed{=}$

C1 Do these calculations, giving each answer correct to 3 significant figures.

(a) $\dfrac{44.2}{3.84 - 1.67}$ (b) $6.73 - (2.95 - 1.08)$ (c) $\dfrac{3.06}{5.13 \times 0.96}$

(d) $\dfrac{12.74 - 8.87}{8.54 + 1.66}$ (e) $\dfrac{14.31 + 10.84}{2.56 \times 1.42}$ (f) $\dfrac{352 - 187}{4.53 - 2.86}$

D Checking by rough estimates

D1 Amber had to calculate $\dfrac{21.39 + 37.78}{4.85}$. She got the answer 29.18, which is wrong.

She should have checked her answer by making a rough estimate:
     21.39 is roughly 20, 37.78 is roughly 40 and 4.85 is roughly 5.

(a) Use these numbers to get a rough estimate of the answer to the calculation.

(b) Do the actual calculation on your calculator.

D2 Pat wants to get a rough estimate for $\dfrac{0.49 \times 216}{3.88}$.

(a) Write down a calculation Pat could do to get a rough estimate.

(b) Work out the rough estimate without using a calculator.

(c) Use a calculator to work out $\dfrac{0.49 \times 216}{3.88}$ and compare the result with your estimate.

D3 For each calculation below

     (i)   work out a rough estimate   (ii)   calculate the result, to 3 significant figures

(a) $\dfrac{57.2}{9.13 - 2.78}$ (b) $4.13 \times (38.5 - 18.8)$ (c) $\dfrac{41.4}{0.97 \times 7.89}$

(d) $\dfrac{207.4 \times 0.48}{28.4 - 9.7}$ (e) $\dfrac{77.31 + 38.84}{5.86 \times 9.75}$ (f) $\dfrac{286 - 18.7}{47.9 + 18.8}$

E *Other keys*

Negative numbers

Most calculators have a 'change sign' key $\boxed{+/-}$ for entering negative numbers.

This key is usually pressed after the number: to enter $^{-}5$ press $\boxed{5}$ $\boxed{+/-}$.

Squaring

The squaring key is often labelled $\boxed{x^2}$. To do 4^2, press $\boxed{4}$ $\boxed{x^2}$.

Square root

On some calculators, the square root key is pressed before the number.

On others it is pressed after. So $\sqrt{9}$ may be $\boxed{\surd}$ $\boxed{9}$ or $\boxed{9}$ $\boxed{\surd}$.

E1 Do each of these first without a calculator.
Then use a calculator to check your answer.
(a) $7 + {}^{-}2$ (b) $^{-}1 + {}^{-}4$ (c) $^{-}7 + 3$ (d) $^{-}5 - {}^{-}5$ (e) $^{-}3 \times {}^{-}4$
(f) $^{-}2 \times 5$ (g) $^{-}10 \div 2$ (h) $^{-}10 \div {}^{-}5$ (i) $3 - {}^{-}1$ (j) $^{-}1 + {}^{-}2.5$

E2 (a) Without a calculator, work out $(^{-}3)^2$.
Then use a calculator to check your answer.
(b) Use a calculator for these.
(i) $(^{-}4.3)^2$ (ii) $3.2^2 - 2.7^2$ (iii) $(7.19 - 4.42)^2$ (iv) $7.19 - 4.42^2$ (v) $(3.2 - 6.7)^2$

E3 Use a calculator to work these out.
Give each answer correct to two decimal places.
(a) $\sqrt{19}$ (b) $6 \times \sqrt{19}$ (c) $\sqrt{6 \times 19}$ (d) $\dfrac{\sqrt{19}}{6}$ (e) $\sqrt{\dfrac{19}{6}}$

E4 Use a calculator for each of these.
Give each answer correct to two decimal places.
Estimate each answer roughly first.
(a) $16.7 - \sqrt{8.91}$ (b) $(2.83 - 1.64)^2$ (c) $4.67 + \left(\dfrac{2.56}{3.2}\right)^2$

(d) $\dfrac{4.94 - 1.8^2}{2.5}$ (e) $3.28 + \dfrac{\sqrt{7.29}}{1.03}$ (f) $\dfrac{4.86}{2.57 - \sqrt{1.69}}$

(g) $\sqrt{4.4^2 + 1.9^2}$ (h) $\dfrac{13.4}{\sqrt{5.76} - 1.54}$ (i) $\sqrt{\dfrac{9.8 + 7.2}{6.8 - 3.5}}$

Test yourself with these questions

T1 Calculate the value of $\dfrac{21.7 \times 32.1}{16.20 - 2.19}$

Give your answer correct to 3 significant figures.   Edexcel

T2 (a) Calculate $\dfrac{89.6 \times 10.3}{19.7 + 9.8}$.

(b) Do not use your calculator in this part of the question.

By using approximations show that your answer to (a) is about right.
You must show all your working.   AQA(SEG)1999

T3 (a) Work out $\dfrac{4.7 \times 20.1}{5.6 - 1.8}$

Write down your full calculator display.

(b) Use estimation to check your answer.
Show each step of your working.   AQA(NEAB)1998

T4 (a) Estimate the value of
$$S = \dfrac{738 \times 19}{593 + 392}.$$

Do not use your calculator.
Show all your approximations and working.

(b) Now use your calculator to work out the value of S.

(i) Write down all the figures in your calculator display.

(ii) Write your answer correct to 3 significant figures.   OCR

T5 Use your calculator to work out the value of $\dfrac{\sqrt{12.3^2 + 7.9}}{1.8 \times 0.17}$

Give your answer correct to 1 decimal place.   Edexcel

T6 (a) Use your calculator to find the value of $3.2^2 - \sqrt{4.84}$.

(b) (i) Use your calculator to find the value of $\dfrac{3.9^2 + 0.53}{3.9 \times 0.53}$.

Write down all the figures on your calculator display.

(ii) Round your answer to part (b) (i) to 2 decimal places.

(iii) Write down a calculation you can do in your head to check
your answer to part (b) (i).

Write down your answer to this calculation.   OCR

T7 Use your calculator to work out the exact value of $\dfrac{14.82 \times (17.4 - 9.25)}{(54.3 + 23.7) \times 3.8}$ Edexcel

Review 4

1 A factory producing computers recorded this information over several weeks.

Week number	1	2	3	4	5	6
Number of computers made	2030	2289	3982	2847	4021	3109
Number of computers faulty	124	109	154	61	62	47

(a) Use percentages to comment on whether the quality of production is improving or getting worse.

(b) The managers plan to produce 3700 computers in week 7 and want at least 99% of them to be fault-free.
How many fault-free computers is that?

2 Use rough estimates to decide
   (a) which of these calculations gives the greatest result
   (b) which gives the smallest result

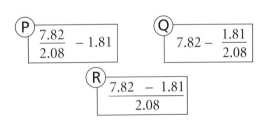

3 For each of these sequences,
   • describe a rule to go from one term to the next
   • find the sixth term
   (a) 2, 9, 16, 23, 30, ...   (b) 0.6, 1.8, 5.4, 16.2, 48.6, ...   (c) 1, 3, 7, 15, 31, ...

4 Work these out on a calculator, giving your answers to 2 decimal places.
   (a) $(5.71 + 3.82) \times 6.7$ (b) $4.45 - 1.82 \times 1.42$ (c) $\dfrac{6.31}{8.22 - 7.46}$

   (d) $\sqrt{7.6 \times (4.2 + 8.1)}$ (e) $(4.2 - 8.9)^2$ (f) $\dfrac{4.3^2 + 0.62}{4.3 + 0.62^2}$

5 The nth term of a series is $\dfrac{12}{n+1}$.

   Calculate the first six terms of the series, correct to 2 decimal places.

6 There are 30 students in a class.
40% of the students are boys.
20% of the students wear glasses.
2 girls wear glasses.

How many boys do not wear glasses?

7 The first term of a sequence of numbers is 1

To find each term of the sequence you multiply the previous term by 4 and then subtract 1.

(a) Write down the first six terms of the sequence.

(b) Work out the first term as a percentage of the second term (to the nearest 1%).

(c) Work out the second term as a percentage of the third term, the third as a percentage of the second, and so on. What do you notice?

8 Matchsticks are used to make these designs.

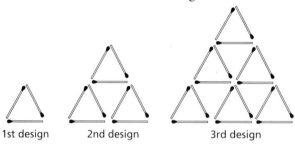

1st design   2nd design   3rd design

(a) Draw the 4th design. How many matchsticks are needed for it?

(b) Is the sequence 3, 9, 18, … linear or non-linear?

(c) Describe a term-to-term rule for the sequence.

(d) How is the sequence 3, 9, 18, … related to the sequence of triangle numbers? What it is about the designs that makes these sequences related in this way? (Use a sketch to help your explanation if you need to).

*9 This metal tray has a square base x cm by x cm.

(a) Write an expression for the area of the base in cm^2.

The metal sides of the tray are 2 cm high.

(b) Write an expression for the area of one side.

(c) Write an expression for the total area of all four sides.

(d) Write an equation for the total area, y cm^2, of metal used for the tray:

$$y = \ldots$$

(e) Make a table with x from 0 to 6 and from it draw a graph of the equation.

(f) Use the graph to find the length of the tray's base if the total area of metal is to be

(i) 40 cm^2 (ii) 25 cm^2 (iii) 80 cm^2

Unitary method

You will revise

◆ using the unitary method to solve problems

You will learn

◆ how to cancel common factors to simplify a calculation

A *Problems*

Prawns with feta cheese

Serves 4

• 2 onions
• 2 large cans chopped tomatoes
• 360 g large peeled prawns
• 100 g feta cheese
• 3 tbsp chopped fresh parsley

What weight of prawns do you need for 6 people?

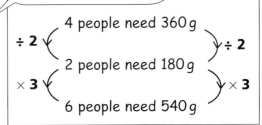

What weight of prawns do you need for 7 people?

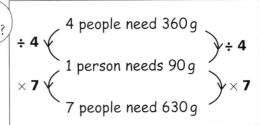

A1 For the recipe above, work these out in your head.

(a) How many onions would you need for 2 people?

(b) How many cans of tomatoes would you need for 6 people?

(c) How much feta cheese would you need for 3 people?

(d) How much parsley would you need for 2 people?

A2 Work these out in your head.

(a) The cost of 3 fruit chews is 12p.
Work out the cost of 5 fruit chews.

(b) The cost of 5 metres of ribbon is £1.50.
Find the cost of 3 metres of the same ribbon.

(c) You need 6 slices of bread for marmalade pudding for 4 people.
How many slices would you need for 6 people?

A3 Here are the ingredients for poached apricots.

(a) What weight of apricots would you need to make poached apricots for 2 people?

(b) How many cardamom pods would you need for 7 people?

(c) How much sugar would you need for 8 people?

(d) What weight of pistachios would you need for 5 people?

> ## Poached apricots
> **Serves 6**
> - 300 g dried apricots
> - 60 ml water
> - 60 g sugar
> - 6 cardamom pods
> - 2 teaspoons of lemon juice
> - 90 g pistachios

A4 Altogether, 3 identical packing cases weigh 240 kg. What will 7 of these packing cases weigh?

A5 Jim lays 8 identical drainage pipes end to end. The total length is 32 m. Work out the total length of 5 of these pipes.

A6 The weight of 7 identical bolts is 140 g. What would 11 of these bolts weigh?

A7 In a recipe for buns, 600 ml of milk is needed to make 20 buns. How much milk would be needed to make 25 of these buns?

A8 The total cost of 3 identical jars of marmalade is £2.94. How much would I pay for 5 jars?

A9 100 g of sugar is needed to make blackberry fool for 5 people. How much sugar would be needed for 12 people?

B *Cancelling common factors*

A $6 \times \dfrac{9}{3}$ B $\dfrac{9 \times 7}{3}$ C 8×3 D $\dfrac{6}{3} \times 9$ E $\dfrac{16}{2} \times 3$ F $\dfrac{6 + 9}{3}$ G $\dfrac{8}{5} \times 15$

H $\dfrac{16}{10} \times 15$ I $\dfrac{9 \times 14}{6}$ J 3×7 K $\dfrac{3 \times 14}{2}$ L $\dfrac{6 \times 9}{3}$ M $6 + \dfrac{9}{3}$

- Find some groups of equivalent expressions.

B1 For each calculation, simplify by cancelling common factors and then evaluate it.

(a) $\dfrac{12 \times 31}{2}$ (b) $\dfrac{25 \times 9}{3}$ (c) $\dfrac{16 \times 27}{4}$

(d) $\dfrac{8 \times 22}{2}$ (e) $\dfrac{21 \times 15}{3}$ (f) $\dfrac{25 \times 15}{5}$

Examples of cancelling common factors

$$\frac{16}{7} \times 14 = \frac{16}{7_1} \times \cancel{14}^{\,2}$$
$$= 16 \times 2$$
$$= 32$$

$$4 \times \frac{13}{8} = {}^1\cancel{4} \times \frac{13}{\cancel{8}\,2}$$
$$= \frac{13}{2}$$
$$= 6.5$$

$$\frac{93 \times 25}{15} = \frac{93 \times \cancel{25}\,5}{\cancel{15}\,3}$$
$$= \frac{{}^{31}\cancel{93} \times \cancel{25}\,5}{\cancel{15}\,\cancel{3}\,1}$$
$$= 31 \times 5$$
$$= 153$$

B2 For each calculation, simplify by cancelling common factors and then evaluate it.

(a) $\dfrac{24}{5} \times 10$

(b) $\dfrac{13}{3} \times 9$

(c) $30 \times \dfrac{13}{6}$

(d) $45 \times \dfrac{8}{15}$

(e) $\dfrac{36}{14} \times 7$

(f) $12 \times \dfrac{7}{24}$

B3 For each calculation, simplify by cancelling common factors and then evaluate it.

(a) $\dfrac{15 \times 26}{6}$

(b) $\dfrac{25 \times 14}{10}$

(c) $\dfrac{22}{4} \times 14$

(d) $25 \times \dfrac{12}{15}$

(e) $\dfrac{21}{12} \times 6$

(f) $\dfrac{27}{18} \times 3$

B4 Joe has 12 bags of sweets, each holding 36 sweets.
He wants to share the sweets equally between 16 children.

(a) Which of these calculations gives the number of sweets each child gets?

$$\frac{36 \times 16}{12} \qquad \frac{12 \times 36}{16} \qquad \frac{12 \times 16}{36}$$

(b) Simplify this calculation by cancelling.
Work out how many sweets each child gets.

B5 Ms Spence has 15 bags of counters.
She wants to share out the counters between the pupils in her class of 25.
Each bag contains 45 counters.

(a) Which of these calculations gives the number of counters each pupil gets?

$$\frac{25 \times 45}{15} \qquad \frac{15 \times 25}{45} \qquad \frac{45 \times 15}{25}$$

(b) Simplify this calculation by cancelling.
Work out how many counters each pupil gets.

B6 Dee has 14 packs of humbugs, each holding 32 humbugs.
She shares the humbugs equally between 28 children.

How many humbugs does each child get?

B7 These toy bricks weigh 30 grams altogether.

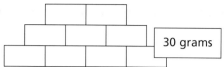

30 grams

(a) Which calculation gives the total weight
of the toy bricks on the right?

$$\frac{30}{9} \times 15 \qquad \frac{30}{15} \times 9 \qquad \frac{15}{9} \times 30$$

(b) Simplify this calculation by cancelling.
Work out the weight of these bricks.

C *Using cancelling*

Example

It takes 130 g of self raising flour
to make 15 chocolate cookies.

How much self raising flour would
you need for 27 chocolate cookies?

15 cookies need 130 g flour

1 cookie needs $\frac{130}{15}$ g flour

27 cookies need $\frac{130}{15} \times 27 = \frac{130^{26}}{15_3} \times 27$

$$= \frac{130^{26}}{15_1} \times 27^{9}$$

$$= 26 \times 9$$

$$= 234 \text{ g flour}$$

C1 12 bolts weigh 40 grams.
How much will 15 of these bolts weigh?

C2 21 nails weigh 28 grams.
How much will 27 of these nails weigh?

C3 It takes 150 ml of milk to make 12 scones.
How much milk would you need for 20 scones?

C4 15 sweets weigh 100 grams.
Find the weight of 21 of these sweets.

C5 It takes 450 g of caster sugar to make 36 marbled fudge bars.
How much caster sugar would you need to make 28 fudge bars?

C6 A tray of 24 cookie twists can be made with 100 g of butter.
How much butter would you need for 42 cookie twists?

C7 You need 90 g of sugar to make 64 Neapolitan cookies.
How much sugar would you need for 48 cookies?

COOKIES

D Using a calculator

Example

A pile of 15 identical books is 38 cm high.

How high will a pile of 35 of these books be?
Give your answer to the nearest cm.

Height of 15 books is   38 cm

Height of 1 book is $\frac{38}{15}$ cm

Height of 35 books is $\frac{38}{15} \times 35$

$= 88.666666 \ldots$

$= 89$ cm (nearest cm)

D1 For 14 hours' work, Lucy was paid £67.06.
At this rate, what should she be paid for 38 hours' work?

D2 A small aircraft flies 275 km on 110 litres of fuel.
How much fuel is needed for a journey of 400 km?

D3 50 ml of milk contains 60 mg of fat.
How much fat is in 568 ml of milk?
Give your answer correct to the nearest mg.

Milk

568 ml

D4 260 sheets of paper have a total thickness of 3.1 cm.
What would be the thickness of 550 of these sheets of paper, to the nearest 0.1 cm?

D5 A Caesar salad recipe for 6 uses 250 g of diced potatoes.
How much diced potato would you need for a Caesar salad for 11 people?
Give your answer correct to the nearest 5 g.

D6 Hayley bought 250 g of olives for £1.89.
What would 160 g of these olives cost?

D7 A shop charges £4.80 for 3.5 m of wire.
At this rate, what would be the cost of 8.2 m of this wire?

D8

BUY £20 000
WORTH OF
PREMIUM BONDS
AND
ON AVERAGE
YOU'LL WIN
13
PRIZES A YEAR.

On average, someone with £20 000 in Premium Bonds should win 13 prizes each year.

Flora has £7 700 in Premium Bonds.

(a) Work out the number of prizes she should win, on average, each year.

Premium Bonds are sold in multiples of £10. Rajesh works out that he is likely to win 9 prizes next year.

(b) Work out an estimate for the amount of money Rajesh has in Premium Bonds.

Edexcel

*D9 A 1 kg block of gold is worth £5 780 and has a volume of 52 cm^3.

> 52 cm^3 is about the size of a medium bar of cholcolate.

    (a) What is the weight of a 65 cm^3 block of gold?

    (b) What weight of gold, to the nearest kilogram, would be worth £1 million?

    (c) What is the volume of this £1 million block, to the nearest cm^3?

E Rates

TG

Example

There are 4.546 litres in one gallon.
Work out the number of gallons in 40 litres, correct to 1 d.p.

4.546 litres is	1 gallon
1 litre is	$\dfrac{1}{4.546}$ gallons
40 litres is	$\dfrac{1}{4.546} \times 40$
	= 8.8 gallons (to 1 d.p.)

4.546 litres is	1 gallon
40 litres is	$\dfrac{40}{4.546}$
	= 8.8 gallons (to 1 d.p.)

E1 The exchange rate to change Norwegian kroners into pounds is

    12.74 Norwegian kroners = £1

Work out the number of pence in 5.50 Norwegian kroners.
Give your answer to the nearest penny.

E2 1 kilogram is equivalent to about 2.205 pounds.
What is the weight in kilograms of 5 pounds of flour?
Give your answer correct to 2 decimal places.

E3 One pint is equivalent to about 0.568 litres.
How many pints are equivalent to 3.6 litres?
Give your answer correct to 1 decimal place.

E4 On average a car travels 100 km on fuel costing £4.40.

    (a) Calculate the cost of fuel for a journey of 720 km.

    (b) The fuel for another journey cost £14.30.
        Calculate the length of this journey.

OCR

E5 A litre is about 0.220 gallons.
A litre of fuel costs 85.9 p.

Calculate the cost of 800 gallons of this fuel.

E6 The table shows the amount of foreign currency that can be bought with £1.

Dave has just returned from Hong Kong.
He has 940 Hong Kong dollars.

His next trip is to South Korea.
He changes his Hong Kong dollars into
South Korean won.

Calculate, to the nearest thousand,
how many South Korean won he will get.

Far-east currencies	
£1 will buy	
India	66.65 rupees
Hong Kong	11.03 Hong Kong dollars
Indonesia	13192.2 rupiahs
South Korea	1816.9 won

Test yourself with these questions

T1 The cost of 3 pencils is 72p.
What is the cost of 5 pencils?   Edexcel

T2 It takes 100 g of flour to make 15 shortbread biscuits.

(a) How many shortbread biscuits can be made from 1 kg of flour?

(b) Calculate the weight of flour needed to make 24 biscuits.

OCR

T3 Recipe for bread and butter pudding.

6 slices of bread
2 eggs
1 pint of milk
150 g raisins
10 g margarine

This recipe is enough for 4 people.

(a) Work out the amounts needed so that there will be enough for 6 people.

There are 450 g in 1 pound. There are 16 ounces in 1 pound.

(b) Change 150 g into ounces. Give your answer correct to the nearest ounce.

Edexcel

T4 A pile of 12 identical coins is 3.8 cm high.

How high will a pile of 25 of these coins be, correct to the nearest mm?

T5 Asra goes on holiday to Hungary.
The exchange rate is £1 = 397.6 forints.

She changes £250 into forints.

(a) How many forints should Asra get?

She changes 1450 forints back into pounds.
The exchange rate is the same.

(b) How much money should she get?
Give your answer to the nearest penny.

22 Similarity

You will revise

◆ how to recognise and make scaled drawings of shapes

You will learn

◆ the properties of similar and congruent shapes

◆ how to use the properties of similar shapes to solve problems

A Scaling

A1 Which of these are scaled copies of the original shape?
What is the scale factor used in each case?

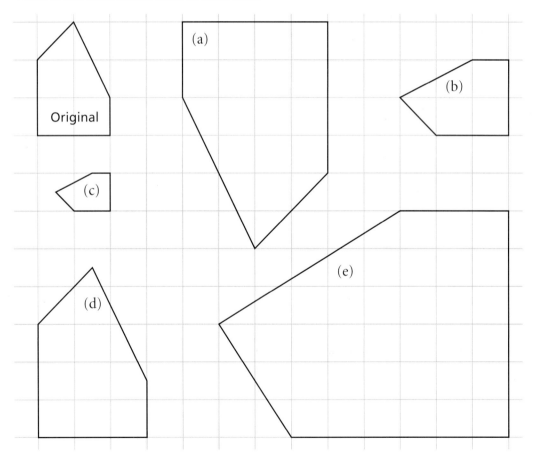

A2 (a) On squared paper copy the original shape above.
Draw a scaled copy of this shape using a scale factor of 4.

(b) Draw another scaled copy using a scale factor of 2.5.

B Scaled copies

A graphic artist has a picture of a church.
He is asked to make a copy but double the size.
He isn't happy about his drawing though.
What is wrong with it?

Original

B1 A designer makes a scaled copy of a drawing of the front of a car using a scale factor of 4.

(a) The height of the car in the original drawing is 5 cm.
What will the height be of the car in the scaled copy?

(b) The thickness of the tyres in the original drawing is 1.5 cm.
What should the thickness of the tyres in the scaled copy be?

(c) The angle made by the windscreen wipers in the original is 140°.
What angle should the windscreen wipers in the scaled copy make?

(d) In the scaled copy the width of the car is 30 cm.
What was the width of the car in the original?

B2 Anna is designing a new shed.
She wants to make a scaled copy
using a scale factor of 2.
Here is her original and her
attempt at a scaled copy.

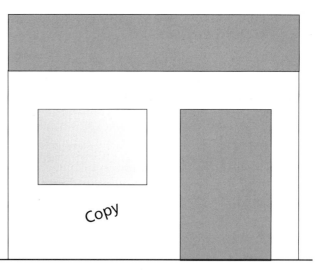

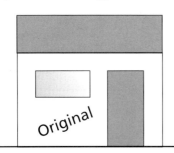

By comparing the measurements on the copy and the original, explain what mistakes
have been made. You may find a table useful.

B3 (a) Measure the length and width of the original kite below.

(b) Measure the length and width of each of the other kites.

(c) Compare each of the kites A,B,C and D with the original.
Which of the kites are true copies of the original?

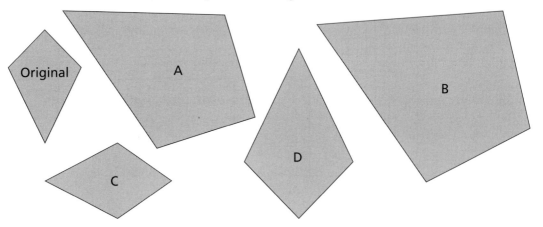

B4 This table compares things in an original picture of a house, car and hedge and a scaled
copy. Copy the table and fill in the missing values.

	Original length	Scale factor	Copy length
Width of picture	10 cm		25 cm
Height of picture	6 cm		
Height of house	3 cm		
Length of car			5 cm
Length of hedge			20 cm

ℂ *Scaling down*

This original pencil is life size.

12 cm

Each pencil below it is $\frac{1}{4}$ of the original size. The scale factor is $\frac{1}{4}$.

3 cm

C1 What scale factor are these copies?

(a)　　　　　　　　　　　　　　　　　　(b)

(c)　　　　　　　　(d)

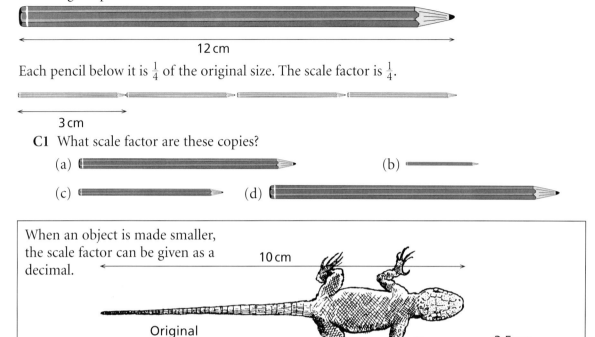

When an object is made smaller,
the scale factor can be given as a
decimal.

10 cm

Original

The scale factor is:
Copy length ÷ Original length
= 3.5 ÷ 10 = 0.35.

3.5 cm

Copy

C2 Find the scale factor for these
copies of the original lizard.

(a)

(b)

(d)

(c)

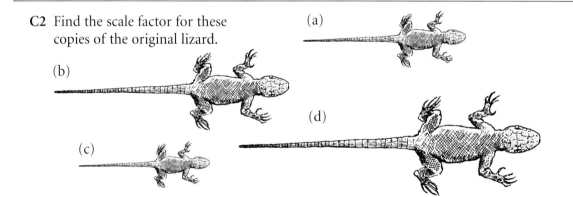

C3 Copy and complete this table for a picture that has been copied.

	Original length	Scale factor	Copy length
Width of picture	8 cm		4.8 cm
Height of picture	5 cm		
Height of tree	3 cm		
Length of pond			4.5 cm
Length of fence			5.4 cm

D Similar triangles

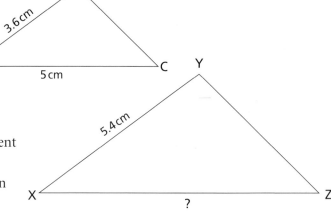

Each of these triangles is an exact scaled copy of the other.

- What is the scale factor of enlargement of ABC to XYZ?

- What is length XZ?

- What is the scale factor of enlargement of length BC to YZ?

- What can you say about the angles in the two triangles

D1 These two triangles are scaled copies of each other.

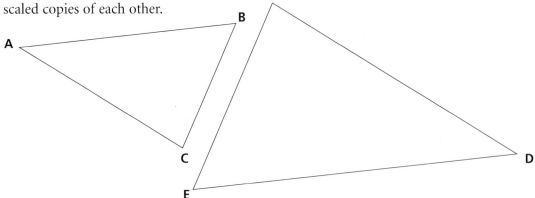

(a) Measure the lengths of AB and DE.
   Use these lengths to find the scale factor of enlargement of DE from AB.

(b) Measure the lengths of BC and EF.
   What is the scale factor of enlargement of BC from EF?

(c) Without measuring give the scale factor of CA to FD.

(d) Measure angle ABC. What can you say about angle DEF?

(e) Measure angle BCA. What can you say about angle EFD?

(f) Without measuring find angles CAB and FDE.

D2 These triangles are all scaled copies of each other.
   Find the missing angles.

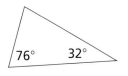
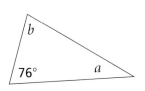
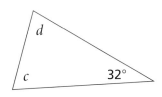

If two triangles are exact scaled copies of each other they are called **similar** triangles.
If the same scale factor is used to enlarge each corresponding
side of two triangles then the triangles are similar.
Any two triangles with all the same angles must be similar.

D3 These triangles are all similar right angled triangles .

(i) From the side given find the scale factor used to enlarge the original triangle.

(ii) Find the missing lengths.

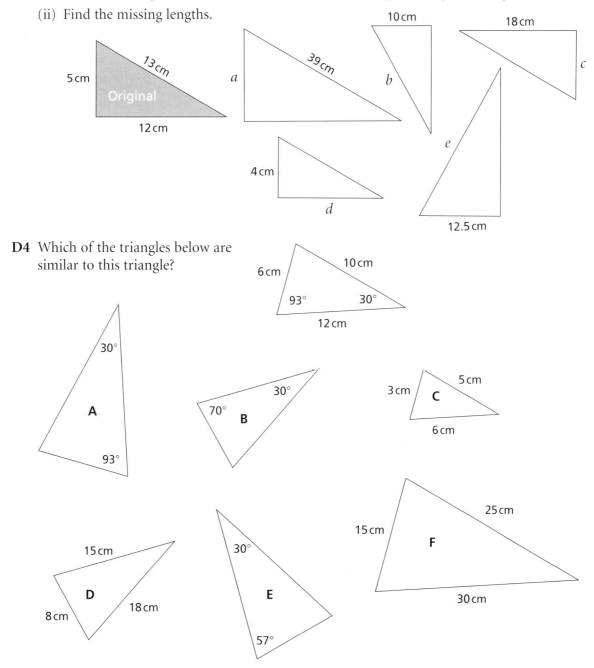

D4 Which of the triangles below are
similar to this triangle?

D5 These triangles are all similar.
Find the missing angles.

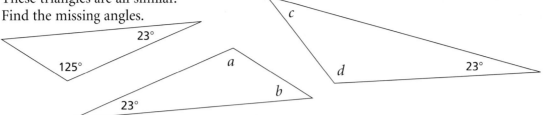

D6 These triangles are similar.

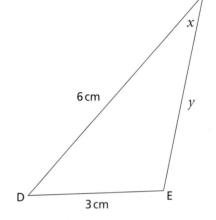

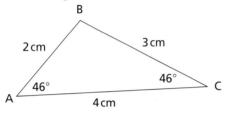

(a) What is the size of angle x?

(b) Calculate the length of y.

AQA(SEG)1997

D7 In this diagram lines AB and DE are parallel.
DE = 7 cm
DC = 4 cm
BC = 6 cm
Angle DEC = 43°

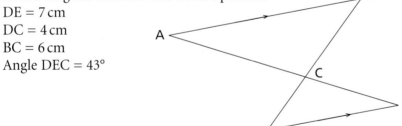

(a) Find angle BAC.

(b) Explain why triangles ABC and DEC must be similar.

(c) What is the scale factor of enlargement of triangle DEC to triangle ABC?

(d) Calculate length AB.

E *Proportions*

Paper sizes

You will need some A2 paper, broadsheet newspaper and a metre rule.

You will often have used 'A4' paper. This is a standard metric size.

Newspapers however use a different system.

Measure the longest and shortest edges of each of your sheets of paper.
Now fold the paper in half. Measure the new longest and shortest edges of the paper.
Repeat this another three times

Record your results in tables:

| | Metric Sizes | | | | | Newspaper Sizes | | | | |
Size	A2	A3	A4	A5	A6	Broadsheet	Tabloid	$\frac{1}{2}$Tab	$\frac{1}{4}$Tab	$\frac{1}{8}$Tab
Longest edge										
Shortest edge										

For a sheet of A2 paper what does the width have to
be multiplied by to obtain the length?

What is this ratio $\frac{\text{longest side}}{\text{shortest side}}$ for other metric paper sizes?

What do you notice?

Work out the ratios $\frac{\text{longest side}}{\text{shortest side}}$ for the newspaper sizes.

Does the same rule apply?

What advantage does the metric sizes system have
over the newspaper sizes?

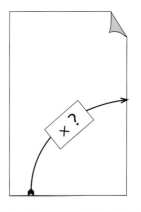

A ratio is used to compare two quantities.

For example if the height of window is 3 times its width,
we can write

$$height = 3 \times width$$

or

$$\frac{height}{width} = 3$$

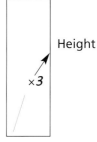

We say 'the ratio of height to width is 3 (or 3:1)'

If two shapes are similar then the ratio between any two sides is the same in both shapes

E1 All the pictures below are scaled copies of this original.

Original

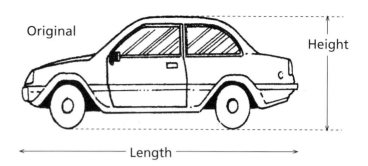

Height

Length

(a) Measure in cm

   (i) the height of the car in the original picture,

   (ii) the length of the car in the original picture.

(b) What is the ratio $\frac{length}{height}$ for this car?

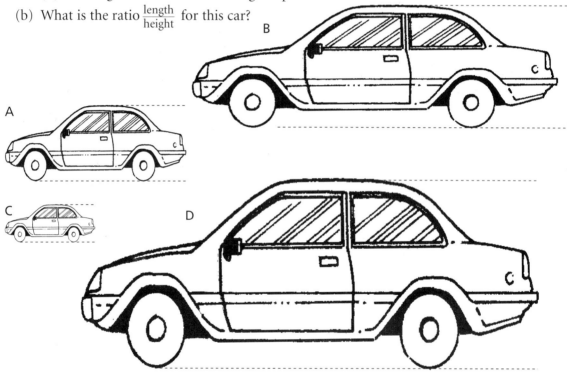

(c) Copy and complete this table;

Copy	A	B	C	D
Length of car				
Height of car				
Ratio $\frac{length}{height}$				

(d) What do you notice about the ratio used each time?

(e) A car is now drawn 7 cm high. How long should it be?

(f) How high would a car 20 cm long be if it were a scaled copy?

E2 For each of these cards find the ratio $\frac{\text{longest side}}{\text{shortest side}}$.
Use this to make pairs which are
copies from the same original

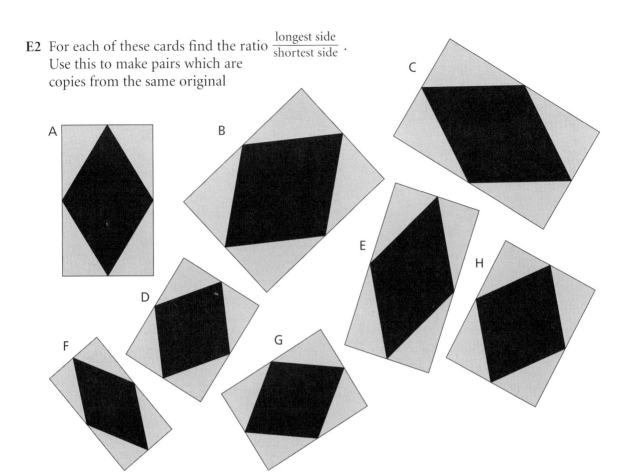

E3 Find the ratio $\frac{x}{y}$ in each of these right-angled triangles.
Use this to decide which are similar to the shaded one.

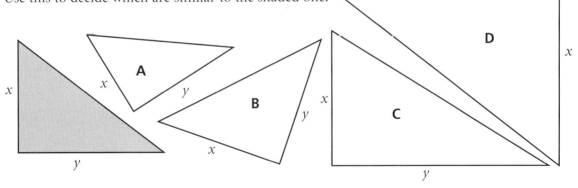

Worked example

The **internal ratio** can be used to find missing sides in similar triangles.
These three triangles are similar.

The internal ratio $\frac{x}{y} = 3 \div 8 = 0.375$.
So $a = 14 \times 0.375 = 5.25$ cm
and $b = 1.8 \div 0.375 = 4.8$ cm

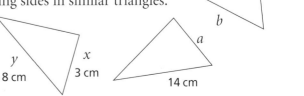

E4 These triangles are all similar to the shaded triangle.

(a) What is the ratio $\frac{XY}{YZ}$ in the original?

(b) Use this ratio to find the lengths of the missing sides.

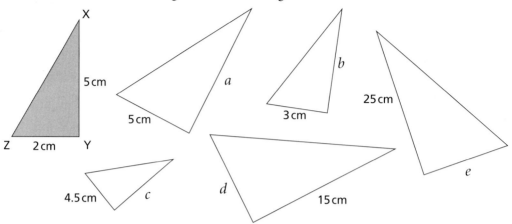

E5 These triangles are all similar to the shaded triangle.

(a) What is the ratio $\frac{NP}{MN}$ in the original?

(b) Use this ratio to find the lengths of the missing sides.

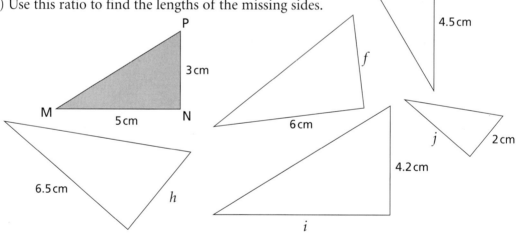

*E6 At a certain time in the afternoon a building which is 12.5 m high casts a shadow which is 10 m long.

At the same time of day another building casts a shadow which is 15 m long.

How high is this building?

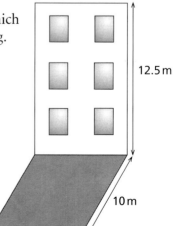

F Solving problems

F1 In this diagram BC is parallel to DE.
AB = 8 cm   BD = 4 cm
DE = 9 cm   Angle ACB = 42°

(a) What is angle CED?

(b) Explain why triangles ABC and ADE must be similar.

(c) What is length AD?

(d) What is the scale factor of enlargement from triangle ABC to ADE?

(e) Find the length BC.

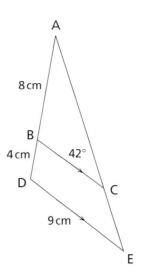

F2 In this diagram PQ is parallel to ST.
PQ = 15 cm   QR = 8 cm
RS = 12 cm   RT = 18 cm

(a) Explain why PQR and RST must be similar triangles.

(b) Calculate length ST.

(c) Calculate length PR.

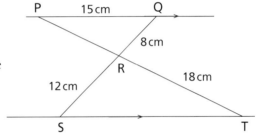

F3 BC is parallel to DE.
AB is twice as long as BD.

AD = 36 cm and AC = 27 cm

(a) Work out the length of AB.

(b) Work out the length of AE

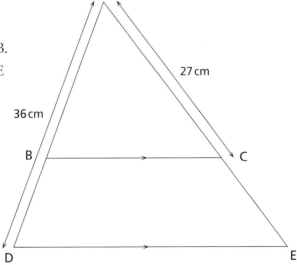

Edexcel

Test yourself with these questions

T1 What scale factor in this diagram has been used to copy

(a) shape A to B

(b) shape C to B

(c) shape A to C?

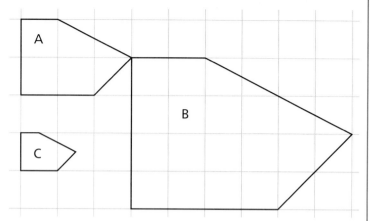

T2 In the diagram, AB is parallel to DE.

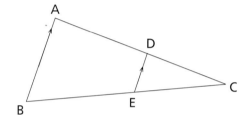

(a) Explain how you know that triangle ABC is similar to triangle DEC.

Length BE is 15 cm and length EC is 9 cm. Length AB is 8 cm.

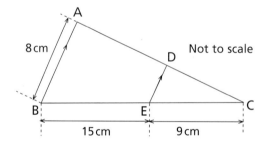

8 cm   Not to scale

15 cm   9 cm

(b) Calculate length DE

AQA(SEG)1998

T3 In the triangle ABC, AB = 10 cm, AC = 24 cm, BC = 26 cm.

Angle BAC = 90°.
D is the midpoint of AB.
DE is perpendicular to BC.

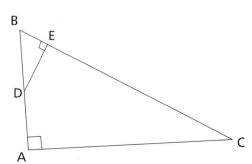

Use similar triangles to calculate the distance DE.

AQA(NEAB)1997

23 Fractions

You will revise

◆ calculating a fraction of a number

You will learn

◆ how to multiply a fraction by an integer (whole number)

◆ how to multiply fractions

A *Fractions review*

Fractions from dice

Roll a dice twice (or roll two dice).
The first score is the numerator of a fraction.
The second is the denominator.

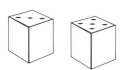

This gives
$\frac{3}{5}$

Roll the dice to make a fraction. Is it

• a proper fraction or an improper fraction?

• equivalent to a whole number?

• a fraction that can be simplified?

• a fraction that gives a recurring decimal?

Repeat for other fractions you can make with the dice.

Challenge!

How many different fractions
do the dice give?

A1 Errol has to calculate three quarters of 52.

He knows he has to do something with 3 and 4 but doesn't know what.
Explain briefly what he has to do.

A2 Calculate these in your head.

(a) $\frac{1}{7}$ of 28   (b) $\frac{2}{7}$ of 28   (c) $\frac{3}{7}$ of 28   (d) $\frac{1}{9}$ of 72   (e) $\frac{4}{9}$ of 72

(f) $\frac{3}{5}$ of 45   (g) $\frac{4}{5}$ of 30   (h) $\frac{3}{8}$ of 32   (i) $\frac{2}{9}$ of 180   (j) $\frac{5}{6}$ of 420

A3 Write each of these fractions in its simplest form.
(Some are already in their simplest form.)

(a) $\frac{20}{30}$ (b) $\frac{12}{20}$ (c) $\frac{18}{40}$ (d) $\frac{15}{40}$ (e) $\frac{12}{35}$

(f) $\frac{16}{21}$ (g) $\frac{18}{21}$ (h) $\frac{24}{60}$ (i) $\frac{14}{42}$ (j) $\frac{75}{360}$

ⓑ *Multiplying a fraction by a whole number*

This picture shows 8 lots of $\frac{1}{2}$.

$$8 \times \tfrac{1}{2} = 4 \qquad \tfrac{1}{2} \times 8 = 4$$

This picture shows 7 lots of $\frac{3}{4}$.

21 quarters = 5 whole ones and 1 quarter

$$7 \times \frac{3}{4} = \frac{21}{4} = 5\tfrac{1}{4}$$

Cancelling common factors

$$8 \times \frac{3}{4} = \overset{2}{\cancel{8}} \times \frac{3}{\cancel{4}_{1}} = 6 \qquad 12 \times \frac{5}{8} = \overset{3}{\cancel{12}} \times \frac{5}{\cancel{8}_{2}} = \frac{15}{2} = 7\tfrac{1}{2}$$

B1 Work these out. The answers are all whole numbers.

    (a) $\frac{1}{4} \times 12$ (b) $15 \times \frac{1}{3}$ (c) $\frac{1}{5} \times 20$ (d) $24 \times \frac{1}{8}$ (e) $32 \times \frac{1}{2}$

B2 Work these out. The answers are all whole numbers.

    (a) $\frac{3}{4} \times 8$ (b) $12 \times \frac{2}{3}$ (c) $\frac{4}{5} \times 10$ (d) $16 \times \frac{1}{8}$ (e) $32 \times \frac{5}{8}$

B3 Janice has 9 glasses of juice.
Each glass contains $\frac{1}{4}$ litre of juice.
How many litres of juice does Janice have?

B4 It takes Prakesh $\frac{3}{4}$ hour to paint a window frame.
He has 5 window frames to paint.
How many hours will it take him to paint them all?

B5 Dylan does a health run of $\frac{2}{3}$ mile every day.
How far does he run altogether in 7 days?

B6 Work these out.

    (a) $\frac{1}{4} \times 15$ (b) $8 \times \frac{1}{3}$ (c) $\frac{1}{5} \times 12$ (d) $14 \times \frac{1}{8}$ (e) $17 \times \frac{1}{2}$

B7 Work these out.

    (a) $\frac{3}{4} \times 9$ (b) $10 \times \frac{2}{3}$ (c) $\frac{4}{5} \times 6$ (d) $6 \times \frac{3}{8}$ (e) $5 \times \frac{4}{5}$

C *'Of' and multiply*

| $\frac{1}{2}$ of 8 = 4 $\frac{1}{2} \times 8 = 4$ | **of** and × give the same result | $\frac{3}{4}$ of 12 = 9 $\frac{3}{4} \times 12 = 9$ |

C1 Copy this: $\frac{\square}{\square}$ of $\square = \square$ Put the digits 3, 4, 6, 8 in the boxes to make the calculation correct.

C2 Put the given digits in the boxes to make each calculation correct.

(a) $\frac{\square}{\square} \times \square = \square$ 2, 3, 4, 6

(b) $\square\square \times \frac{\square}{\square} = \square$ 0, 1, 2, 4, 5

(c) $\frac{\square}{\square}$ of $\square\square = \square$ 0, 1, 4, 5, 8

(d) $\frac{\square}{\square} \times \square\square = \square\square$ 0, 1, 2, 3, 4, 5

C3 Use the fact that $\frac{2}{3}$ of 8 = $\frac{2}{3} \times 8$ to work out $\frac{2}{3}$ of 8. Write the answer as a mixed number.

C4 Work out   (a) $\frac{1}{4}$ of 7   (b) $\frac{3}{4}$ of 15   (c) $\frac{2}{3}$ of 10   (d) $\frac{2}{5}$ of 14

D *Dividing a fraction by a whole number*

This is $\frac{1}{2}$ 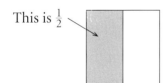 and this is $\frac{1}{2} \div 3$ 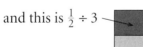 What fraction is $\frac{1}{2} \div 3$

This is $\frac{2}{3}$ 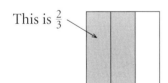 and this is $\frac{2}{3} \div 5$ What fraction is $\frac{2}{3} \div 3$

D1 Work these out.

(a) $\frac{1}{4} \div 2$ (b) $\frac{1}{3} \div 2$ (c) $\frac{1}{3} \div 3$ (d) $\frac{1}{2} \div 5$ (e) $\frac{1}{5} \div 4$

D2 Work these out.

(a) $\frac{3}{4} \div 2$ (b) $\frac{2}{3} \div 2$ (c) $\frac{2}{5} \div 3$ (d) $\frac{3}{4} \div 5$ (e) $\frac{4}{5} \div 4$

D3 Work these out.

(a) $\frac{3}{5} \times 3$ (b) $\frac{2}{3} \div 5$ (c) $\frac{2}{5} \times 7$ (d) $\frac{3}{8} \div 2$ (e) $\frac{4}{5} \times 6$

E Fractions of fractions

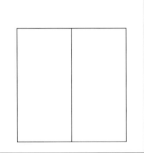

This square represents 1 unit

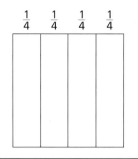

First it is split vertically into **quarters.**

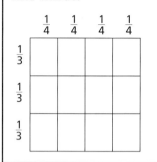

Then split horizontally into **thirds.**

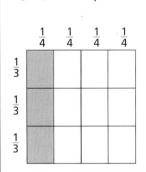

Lightly shade $\frac{1}{4}$.

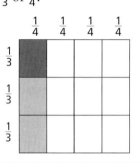

Then heavily shade $\frac{1}{3}$ of $\frac{1}{4}$.

What fraction of the square is $\frac{1}{3}$ of $\frac{1}{4}$?

• What does each of these diagrams show?

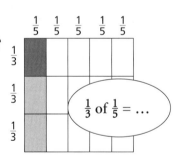

$\frac{1}{3}$ of $\frac{1}{5}$ = ...

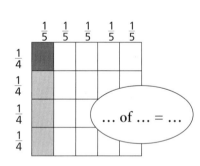

... of ... = ...

E1 What does each of these diagrams show?

(a)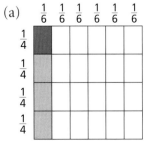

(b)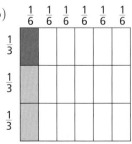

(c)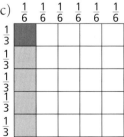

E2 "I lightly shade $\frac{3}{4}$, and heavily shade $\frac{2}{3}$ of $\frac{3}{4}$."

$\frac{3}{4}$

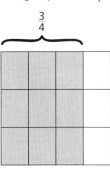

$\frac{3}{4}$

$\frac{2}{3}$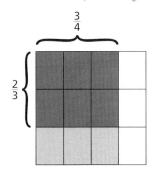

What fraction of the square is $\frac{2}{3}$ of $\frac{3}{4}$?

E3 "I lightly shade $\frac{3}{5}$, and heavily shade $\frac{3}{4}$ of $\frac{3}{5}$."

$\frac{3}{5}$

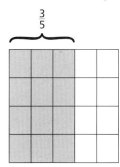

$\frac{3}{5}$

$\frac{3}{4}$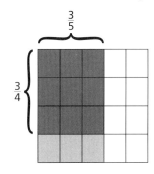

What fraction of the square is $\frac{3}{4}$ of $\frac{3}{5}$?

E4 What does each of these diagrams show?

(a)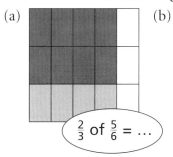

$\frac{2}{3}$ of $\frac{5}{6}$ = …

(b)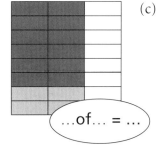

…of… = …

(c)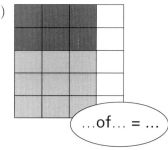

…of… = …

E5 Draw this diagram.

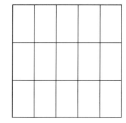

By light and dark shading, show some different fractions of fractions (at least three).

Here is one to start you off.

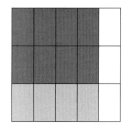

$\frac{1}{3}$ of $\frac{4}{5}$ = …

𝔽 *Multiplying fractions*

$\frac{1}{2} \times \frac{1}{4}$ and $\frac{1}{2}$ of $\frac{1}{4}$ mean the same.

$\frac{1}{2} \times \frac{1}{4} = \frac{1}{8}$

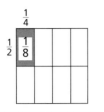

This diagram shows that $\frac{2}{3} \times \frac{4}{5} = \frac{8}{15}$

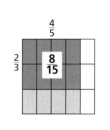

Notice that to multiply fractions, you multiply the numerators and multiply the denominators.

$$\frac{2}{3} \times \frac{4}{5} = \frac{8}{15}$$

F1 Work these out.

(a) $\frac{1}{2} \times \frac{1}{3}$ (b) $\frac{1}{3} \times \frac{1}{4}$ (c) $\frac{2}{3} \times \frac{1}{5}$ (d) $\frac{2}{3} \times \frac{4}{5}$ (e) $\frac{3}{4} \times \frac{3}{5}$

F2 Work these out. Simplify the result where possible.
(The first is done for you as an example.)

(a) $\frac{4}{5} \times \frac{3}{8} = \frac{12}{40} = \frac{3}{10}$ (b) $\frac{2}{3} \times \frac{1}{4}$ (c) $\frac{3}{4} \times \frac{5}{6}$ (d) $\frac{2}{3} \times \frac{3}{8}$

Question F2 (a) can also be done by **cancelling** common factors first: $\dfrac{\overset{1}{\cancel{4}}}{5} \times \dfrac{3}{\underset{2}{\cancel{8}}} = \dfrac{3}{10}$

F3 Work these out, giving each result in its simplest form.

(a) $\frac{1}{3} \times \frac{3}{4}$ (b) $\frac{2}{3} \times \frac{5}{6}$ (c) $\frac{2}{3} \times \frac{5}{8}$ (d) $\frac{4}{5} \times \frac{7}{8}$ (e) $\frac{4}{5} \times \frac{5}{8}$

F4 Work these out, giving each result in its simplest form.

(a) $\frac{1}{8} \times \frac{4}{5}$ (b) $\frac{5}{8} \times \frac{3}{10}$ (c) $\frac{4}{5} \times \frac{7}{10}$ (d) $\frac{2}{3} \times \frac{6}{7}$ (e) $\frac{7}{8} \times \frac{5}{6}$

F5 Work these out.

(a) $2 \times \frac{1}{2}$ (b) $\frac{1}{3} \times 3$ (c) $5 \times \frac{1}{5}$ (d) $\frac{2}{3} \times \frac{3}{2}$ (e) $\frac{3}{4} \times \frac{4}{3}$

Mixed numbers: worked example

Work out $2\frac{2}{3} \times \frac{3}{4}$.

Change the mixed number to an improper fraction.

$2\frac{2}{3} = 2 + \frac{2}{3} = \frac{6}{3} + \frac{2}{3} = \frac{8}{3}$. So $2\frac{2}{3} \times \frac{3}{4} =$

F6 Work out

(a) $2\frac{1}{2} \times \frac{1}{3}$ (b) $\frac{1}{4} \times 3\frac{1}{2}$ (c) $2\frac{1}{3} \times \frac{2}{3}$ (d) $\frac{3}{4} \times 1\frac{1}{5}$

F7 Work out

(a) $1\frac{1}{2} \times 1\frac{1}{3}$ (b) $1\frac{1}{4} \times 2\frac{1}{2}$ (c) $2\frac{1}{3} \times 1\frac{2}{3}$ (d) $1\frac{2}{3} \times 2\frac{1}{2}$

Ⓖ *Mixed questions*

These questions cover all the work on fractions done so far.

G1 (a) What numbers are missing here? (i) $\frac{1}{4} = \frac{}{12}$ (ii) $\frac{1}{3} = \frac{}{12}$

(b) Use the answers to (a) to work out $\frac{1}{4} + \frac{1}{3}$. (c) Work out $1\frac{1}{4} + 2\frac{1}{3}$.

(d) Work out $\frac{1}{3} - \frac{1}{4}$. (e) Work out $2\frac{1}{3} - 1\frac{1}{4}$. (f) Work out $3\frac{1}{4} - 1\frac{1}{3}$.

G2 Work out (a) $\frac{1}{2} + \frac{1}{3}$ (b) $\frac{3}{4} + \frac{1}{8}$ (c) $\frac{5}{6} - \frac{1}{4}$ (d) $\frac{3}{5} + \frac{3}{4}$

G3 Marty has a jug which holds $1\frac{1}{4}$ litres. He finds that 7 jugfuls
will just fill his fish tank. How many litres does his fish tank hold?

G4 Work these out.
(a) $9 \times \frac{3}{4}$ (b) $\frac{1}{3}$ of 20 (c) $\frac{2}{3} \times 10$ (d) $\frac{3}{4}$ of 22 (e) $7 \times \frac{2}{5}$

G5 Work these out.
(a) $\frac{1}{3} \div 2$ (b) $\frac{1}{2} \div 4$ (c) $\frac{2}{3} \div 3$ (d) $\frac{3}{4} \div 3$ (e) $\frac{3}{5} \div 6$

G6 Work these out.
(a) $\frac{1}{3} \times \frac{3}{4}$ (b) $\frac{1}{2} \times \frac{2}{3}$ (c) $\frac{2}{3} \times \frac{3}{4}$ (d) $\frac{3}{4} \times \frac{3}{4}$ (e) $\frac{3}{5} \times \frac{2}{3}$

G7 Work these out.
(a) $1\frac{1}{3} + \frac{3}{4}$ (b) $1\frac{1}{5} \times \frac{2}{3}$ (c) $\frac{5}{6} - \frac{2}{3}$ (d) $\frac{3}{4} \times 1\frac{3}{8}$ (e) $1\frac{3}{5} + 2\frac{2}{3}$

Test yourself with these questions

T1 Jim said 'I've got three quarters of a tin of paint".
Mary said 'I've got four sixths of a tin of paint and my tin
of paint is the same size as yours'.

Who has the most paint, Mary or Jim? Explain your answer.

Edexcel

T2 Two rods are fastened together.
The total length is $3\frac{1}{3}$ inches.
The length of rod B is $1\frac{3}{4}$ inches.
Find the length of rod A.

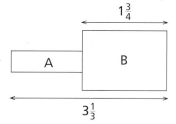

Edexcel

T3 Work out the following.
Give your answers as simply as possible.
(a) $\frac{2}{3} + \frac{4}{5}$ (b) $2\frac{2}{5} \times \frac{5}{6}$

OCR

24 Speed, distance, time

Before you start, you need to be able to do problems like this one:
'If you need 120 g of flour to make 4 buns, how much do you need to make 15 buns?'

From the work in this unit you should learn how to

◆ calculate speeds, distances and times

◆ read and draw travel graphs

A Speed

Gonzalez   24 cm in 4 minutes

Blitz

35 cm in 7 minutes

Which snail is fastest?
Which is slowest?

33 cm in 5 minutes

Nijinsky

39 cm in
6 minutes

A1 A cyclist going at a constant speed travels 48 metres in 6 seconds.
Calculate her speed in metres per second (m/s).

A2 Four pigeons are released together.
These are the distances they fly and the times taken.

Pigeon	A	B	C	D
Distance flown (km)	100	90	174	144
Time taken (hours)	4	3	6	3

Calculate the average speed of each pigeon in km per hour (km/h).

A3 Calculate the average speed of each of the following.
Write the units as part of your answer (for example, km/h).

(a) A train that goes 140 km in 2 hours

(b) A man who runs 150 m in 15 seconds

(c) An aircraft that takes 3 hours to fly 840 miles

(d) A coach that travels 20 miles in half an hour

Rocket

Worked example

What is the average speed, in m.p.h., of a ship which travels 15 miles in 2.5 hours?

First method

15 miles in 2.5 hours

= 30 miles in 5 hours (doubling)

= 6 miles in 1 hour (dividing by 5)

So speed = **6 m.p.h.**

Second method

speed in m.p.h. = distance in miles / time in hours

= $\frac{15}{2.5}$ = $\frac{30}{5}$ = **6 m.p.h.**

A4 Calculate the average speed of each of these, stating the units.

(a) A car that goes 65 miles in 2 hours

(b) A ship that takes 5 hours to sail 75 km

(c) A plane that flies 210 km in $\frac{1}{2}$ hour

(d) A horse that runs 300 m in 20 seconds

A5 A ferry crosses an estuary, which is 18 miles wide, in $1\frac{1}{2}$ hours.
Calculate its average speed.

A6 A plane flying at a constant speed flies 200 miles in $1\frac{1}{4}$ hours.

(a) How far would it fly in $2\frac{1}{2}$ hours?

(b) How far in 5 hours at this speed?

(c) What is the speed of the plane, in m.p.h.?

A7 A coach takes $3\frac{1}{2}$ hours to travel from Hull to Birmingham, a distance of 140 miles.
Calculate the average speed of the coach.

A8 At the start of a journey, the mileometer on Sharmila's car reads 24 752 miles.
At the end of the journey the mileometer reads 24 941 miles.
The journey took $4\frac{1}{2}$ hours.

Calculate the average speed for the journey.

A9 A non-stop flight of 5625 miles takes 12.5 hours.
Calculate the average speed of the aircraft.

*A10 A train leaves London at 08:15 and travels non-stop to York, arriving at 10:00.
The distance from London to York is 189 miles.

(a) Work out the time taken for the journey, in hours.

(b) Calculate the average speed of the train.

B Travel graphs

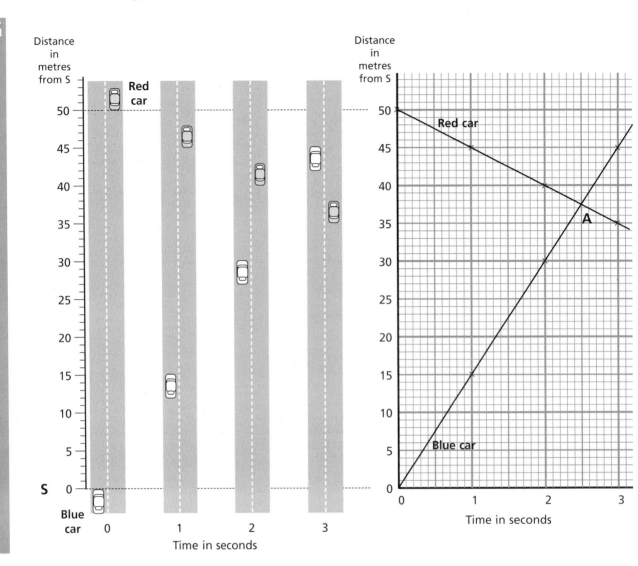

1 What is the speed of the blue car?

2 What is the speed of the red car?

3 What does the point A on the graph show?

4 How far apart are the cars after 1.5 seconds?

5 How far apart are they after 3 seconds?

6 When will the red car reach the point where the blue car was to start with?

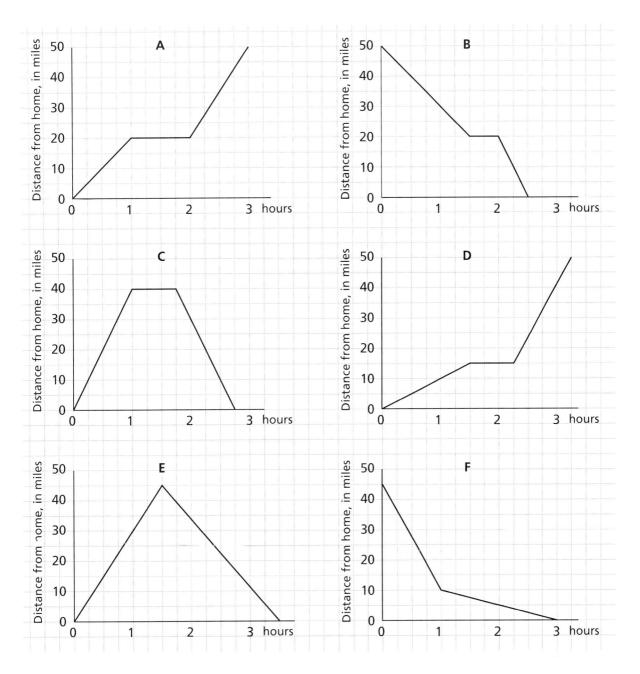

B1 Describe each journey fully.

For example, graph A:

Stage 1: 20 miles in 1 hour = 20 mph

Stage 2: stopped for 1 hour

Stage 3:

B2 Edward catches a bus from Ashbridge to Benmouth.
Julie does the same journey by car.
This graph shows their journeys.

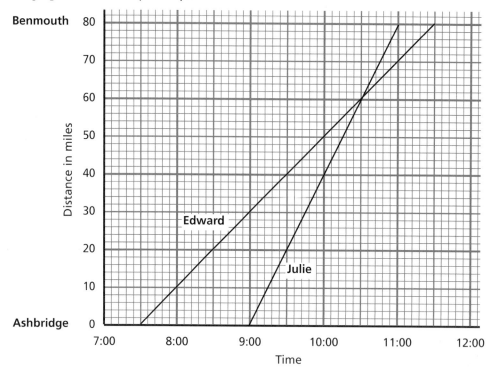

(a) How far is it from Ashbridge to Benmouth?

(b) At what time does Edward leave Ashbridge?

(c) How far apart are Edward and Julie at 9:30?

(d) When does Julie overtake Edward?

(e) What is the speed of (i) the car (ii) the bus

B3 Make a copy on graph paper of the axes above.

(a) Roger leaves Ashbridge at 8:00 and travels to Benmouth at 30 m.p.h.
Draw and label the graph of his journey.

(b) Sally leaves Benmouth at 8:00 and travels to Ashbridge at 20 m.p.h.
Draw and label the graph of her journey.

(c) Roger and Sally pass each other.
How far are they from Ashbridge when they pass?

(d) At what time do they pass?
(First work out how many minutes one small grid square stands for.)

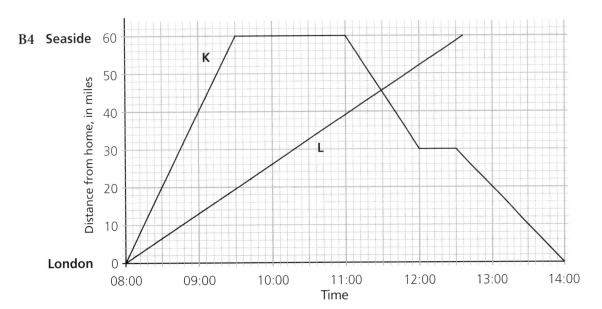

B4 Kylie drives from London to the seaside and back. Graph K shows her journey.

(a) At what time did Kylie leave London?

(b) How long did it take her to reach the seaside?

(c) At what speed did she travel to the seaside?

(d) How long did she spend at the seaside?

(e) On the way home, Kylie stopped for a meal. For how long did she stop?

(f) At what speed was she travelling before she stopped for a meal?

(g) What was her speed for the last part of the journey home?

Kylie's brother Lee drives an old car. Graph L shows his journey to the seaside.

(h) At what time did Kylie and Lee pass each other?

(i) How far were they from London when this happened?

(j) At what time did Lee get to the seaside?

B5 Make a copy of the axes above on graph paper.

(a) Rajesh leaves London at 8:00 and cycles towards the seaside.
He travels at 10 m.p.h. for 2 hours and stops for an hour's rest.
He decides to return and arrives home at 14:00.

Draw and label the graph of his journey.

(b) Nina leaves the seaside at 11:00 and drives towards London at 40 m.p.h.
Draw and label the graph of her journey.

(c) At what time does Nina pass the place where Rajesh stopped for a rest?

(d) At what time does Nina overtake Rajesh?

C Calculating distances

A car travelling at 12 metres per second goes …

$12 \times 1 = \mathbf{12\ metres}$ in 1 second

$12 \times 3 = \mathbf{36\ metres}$ in 3 seconds

Speed	×	time	=	distance
in m/s		in seconds		in metres

C1 Patrick rides his bike at 6 m/s for 15 seconds. How far does he travel?

C2 Henry flies his plane at 240 m.p.h. for 3 hours. How far does he travel?

C3 A slug crawls across a path at a speed of 0.5 cm per second.
How far does it move in

    (a) 10 seconds   (b) 25 seconds   (c) 1 **minute**   (d) $2\frac{1}{2}$ minutes

C4 A coach leaves London at 4:00 p.m. and travels at 55 m.p.h.
How far has it travelled at 7:00 p.m.?

C5 A ferry which travels at 12 m.p.h. leaves at 12:30 and arrives at 14:00.
How far does it travel in this time?

C6 A plane flies at 250 m.p.h.
How far does it travel in   (a) 4 hours   (b) $4\frac{1}{2}$ hours   (c) 7 hours

D Mixing units

Worked example

A train travels 12 miles in 6 minutes.
Calculate its average speed in m.p.h.

First method

To get from 6 minutes to 60 minutes,
I must multiply by 10.

  12 miles in 6 minutes

  12 × 10 miles in 60 minutes

  120 miles per hour

Second method

In 6 minutes it goes 12 miles.

So in 1 minute it goes $\frac{12}{6}$ = 2 miles.

So in 60 minutes it goes 2 × 60 = 120 miles,

Speed = **120 m.p.h.**

D1 A car travels 8 miles in 15 minutes.
Calculate its average speed in miles per hour.

D2 A boat takes 5 minutes to travel 2 miles. Calculate its speed in m.p.h.

D3 A horse runs at a speed of 10 metres per second.
How far does it run in   (a) 1 minute   (b) 5 minutes   (c) 15 minutes

D4 A jet fighter plane flies at 600 m.p.h.
How far does it travel in (a) 15 minutes (b) 1 minute   (c) 30 seconds

D5 Carla drove 3.5 miles in 10 minutes. Work out her average speed in m.p.h.

D6 If your speed is 30 kilometres per hour, how far do you travel in
(a) 20 minutes   (b) 5 minutes   (c) 40 minutes   (d) 45 minutes

E *Hours and minutes on a calculator*

When you use these formulas, …

$$\text{speed} = \frac{\text{distance}}{\text{time}}$$

$$\text{distance} = \text{speed} \times \text{time}$$

… you must be careful about **units**.

distance	time	speed
miles	hours	miles per hour
metres	seconds	metres per second
kilometres	minutes	kilometres per minute

Worked example

A ship travels 37 miles in 1 hour 25 minutes. Calculate its average speed in m.p.h.

First change 25 minutes to **hours** by dividing by 60: $\frac{25}{60} = 0.41666\ldots$

So 1 hour 25 minutes = 1.41666… hours

Average speed = $\frac{\text{distance}}{\text{time}}$ = $\frac{37}{1.41666\ldots}$ = **26.1 m.p.h.** (to 1 decimal place)

Give answers to these questions to one decimal place.

E1 Calculate the average speed of a cyclist who covers 26 miles in 1 hour 35 minutes.

E2 Find the average speed of a swimmer who takes 2 hours 10 minutes to swim 11 miles.

E3 A long distance walker walks for 1 hour 12 minutes at a constant speed of 5.5 m.p.h.
How far does he walk in this time?

E4 A coach travels a distance of 25 miles in 35 minutes.
Calculate the average speed of the coach in m.p.h., to one decimal place.

E5 Calculate the average speed, in m.p.h., of a cyclist who covers 12 miles in 50 minutes.

E6 A train is travelling at 75 m.p.h. How far does it go in 25 minutes?

F *Calculating times*

Worked example

A canal boat travels at 5 kilometres per hour.
How long does it take to travel 30 kilometres?

First method

Imagine the distance the boat has to travel.

_____
                                    30 km

In each hour it travels 5 km.

1 hour
➤➤➤➤ ————————
5 km

So find how many 5s are in 30: $\dfrac{30}{5}$ = **6 hours**

Second method

First find how long it takes to travel **1 km**.

5 km take 1 hour.

So 1 km takes $\dfrac{1}{5}$ hour.

So 30 km take $30 \times \dfrac{1}{5}$ = **6 hours**

Notice that $\text{time} = \dfrac{\text{distance}}{\text{speed}}$

F1 Pam walks at a steady speed of 4 km/h. How long does she take to walk 12 km?

F2 Alvin's speedboat travels at 20 m.p.h. How long does it take to travel 80 miles?

F3 Jim drives at a steady speed of 25 m.p.h. How long does he take to travel 150 miles?

F4 How long does it take to walk 5 miles at a steady speed of 2 m.p.h.?

F5 A boat is 6 miles from the coast and drifting at $1\frac{1}{2}$ m.p.h. towards the coast.
How long will it take the boat to reach the coast?

F6 How long does it take to travel 100 miles at a steady speed of 40 m.p.h.?

Using a calculator

Worked example

A lifeboat travels at 35 kilometres per hour.
How long does it take to reach a ship in danger 48 kilometres away?

Use the formula   time = $\dfrac{\text{distance}}{\text{speed}}$.   Time taken = $\dfrac{48}{35}$ = **1.3714... hours**.

Change the decimal of an hour to minutes by multiplying by 60:
    0.3714... × 60 = **22 minutes** (to nearest minute)

So the time taken is **1 hour 22 minutes**.

F7 A train travels at 55 m.p.h.
How long, in hours and minutes, does it take to travel 128 miles?

F8 A motorway coach travels at a steady speed of 85 km/h.
How long, in hours and minutes, does it take to travel 205 km?

F9 The distance between two Channel ports is 42 km.
How long will the journey take

(a) in a ferry travelling at 18 km/h   (b) in a hydrofoil travelling at 50 km/h

F10 A plane flies a distance of 254 miles at a speed of 320 m.p.h.
How long, in minutes, does the journey take? Give your answer to the nearest minute.

Ⓖ *Mixed questions*

G1 Calculate the missing entries
in this table.

Distance	Time	Average speed
50 miles	4 hours	(a)
(b)	3 hours 30 minutes	42 m.p.h.
64 km	(c)	40 km/h
48 km	1 hour 15 minutes	(d)
35 miles	(e)	50 m.p.h.

G2 A lift in a tall office block travels non-stop between the ground floor and the 30th floor, a distance of 92 metres.

The lift travels at 5 metres per second.
How long does the journey take, to the nearest second?

G3 Boris ran 800 metres in 1 minute 45 seconds.
Calculate his average speed in metres per second, to one decimal place.

Test yourself with these questions

T1 (a) A train takes 3 hours to travel 150 miles.
What is its average speed?

(b) Another train travels 50 miles at an average speed of 37.5 m.p.h.
How long does the journey take?
Give your answer in hours and minutes.

AQA(NEAB)1998

T2 The travel graph shows the journey of a cyclist from the town of Selby.

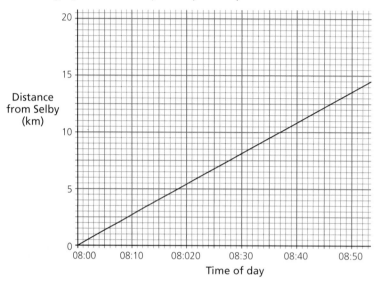

(a) What is the average speed of the cyclist in kilometres per hour?

A motorist is driving towards Selby along the same road as the cyclist. At 08:20 the motorist is 20 km from Selby and travelling at a uniform speed of 60 km/h.

(b) (i) Copy the graph above and draw a second graph on the same axes to show the journey of the motorist to Selby.

(ii) At what time does the motorist pass the cyclist?

AQA(SEG)2000

T3 In the College Games, Michael Jackson won the 200 metres race in a time of 20.32 seconds.
Calculate his average speed in metres per second. Give your answer correct to 1 decimal place.

Edexcel

T4 (a) The train from London to Manchester takes 2 hours 30 minutes.
this train travels at an average speed of 80 miles per hour.
What is the distance from London to Manchester?

(b) The railway company is going to buy some faster trains.
These new trains will have an average speed of 100 miles per hour.
How much time will be saved on the journey from London to Manchester?

AQA(NEAB)1998

Review 5

1 Work these out

(a) $\frac{1}{7} \times 35$ (b) $48 \times \frac{1}{4}$ (c) $\frac{4}{5} \times 35$ (d) $84 \times \frac{3}{4}$ (e) $\frac{1}{4} \times 11$

(f) $17 \times \frac{2}{3}$ (g) $\frac{1}{6} \div 4$ (h) $\frac{2}{3} \div 5$ (i) $\frac{3}{5} \times \frac{5}{6}$ (j) $\frac{1}{3} + \frac{1}{5}$

2 Joan is paid £117.88 for 14 hours' work.
How much will she be paid for 25 hours' work at the same rate?

3 This key is drawn to its exact size.

What scale factor has been used to produce each of these copies?

(a)

(b)

4 A cat eats $\frac{2}{3}$ of a can of cat food each day.

(a) How many cans are needed to feed the cat for 12 days?

(b) How long will 14 cans of cat food last?

5 The instructions for some soluble lawn food say
'450 grams of lawn food is enough to feed $100\,\text{m}^2$ of lawn'.
How much lawn food is needed for a rectangular lawn 18 m by 15 m ?

6 Draw a scalene triangle.
Divide it into four congruent triangles that are similar to it.

7 At the start of a journey a car's mileometer showed 042786.
At the end it showed 042936.
The journey took $3\frac{3}{4}$ hours.

(a) Calculate the average speed in m.p.h.

(b) Given that 1 mile is approximately 1.6 kilometres,
convert the speed to kilometres per hour.

8 A row of 11 one-pound coins is 24.6 cm long.
How long is a row of 7 one-pound coins (to the nearest 0.1 cm)?

9 A photographic shop offers these sizes of colour prints.

Are any of the rectangles similar to one another?
Explain how you decided.

100 mm	by	150 mm
125 mm	by	175 mm
150 mm	by	225 mm
175 mm	by	250 mm

10 (a) Explain why triangle PTS and triangle PRQ are similar triangles.

(b) Calculate the length QR.

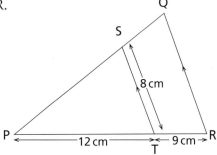

11 This graph shows Karen's journey to school.
She walks to a bus stop, waits for a bus, then
a bus takes her the rest of the way to school.

(a) How far does Karen walk from home to
the bus stop?

(b) At what speed does she walk, in m.p.h.?

(c) How long does she wait at the bus stop?

(d) How far does the bus take her?

(e) At what speed does the bus go, in m.p.h.?

(f) Karen's brother leaves home 12 minutes
after she does. At what speed must he run
in order to catch the same bus as her?

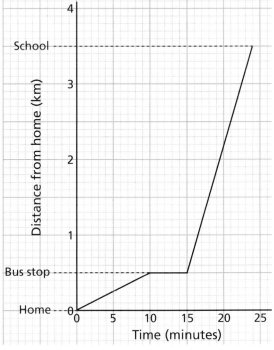

12 A pint is approximately 0.57 of a litre.
A supermarket sells milk at 78 p for a 2-pint carton.
How much is this per litre?

13 A train travels at 65 m.p.h.
How long, in hours and minutes does it take take to travel 148 miles?

25 Constructions

For this work you will need

◆ to be able to construct an equilateral triangle using a pair of compasses only

◆ to be familiar with different types of triangles and quadrilaterals

You will learn how to carry out and use 'constructions' (drawing methods that use a ruler and a pair of compasses but not a set square or an angle measurer)

A The shortest route from a point to a line

The shortest route from a point to a line is at right angles to the line.

This seems obvious when the line is horizontal … … or vertical.

It is less obvious (but still true) when the line is at some other angle.

'Constructing' a perpendicular from a point to a line

1 Suppose you have been given a point P and a line *l*.
Draw an arc with its centre at P, crossing line *l*.

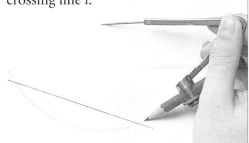

2 Put the point of the compasses at one of the points where your arc crosses line *l*.

Draw an arc below the line.
You need not use the same radius as before.

3 Do not alter the radius.
Draw another arc like this.

4 The line from P to where the last pair of arcs cross is perpendicular to line *l*.

Can you explain why?

A1 Use the construction shown on the previous page
to answer the questions on sheet P54.

A2 Construct an equilateral triangle with sides 15 cm long.
Mark a point P anywhere inside the triangle.

Construct perpendicular lines from P to
each side of the triangle.

Measure the distance from P to each side of
the triangle. Add the three distances together.

Compare your result with at least two
neighbours' results. What do you find?

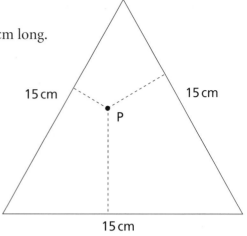

15 cm   15 cm

15 cm

Triangle investigation 1

In the middle of a piece of A4 paper, draw a scalene triangle with sides between 8 and 16 cm
long and all its angles acute. Label its vertices A, B and C.

Using a ruler and compasses, construct

- a line from A perpendicular to BC
- a line from B perpendicular to AC
- a line from C perpendicular to AB

What do you notice about these three lines?
Did what you notice happen with other people's triangles?

What happens when you try to do this for a triangle with one obtuse angle?

Ⓑ *The bisector of an angle*

Bisecting means cutting in half. Follow this construction to bisect an angle.

1 Draw an arc with its centre at the vertex of the angle.	**2** Draw two arcs with the same radius from the points where your first arc crosses the arms of the angle.	**3** Draw the line that bisects the angle. (Can you explain why it bisects the angle?)
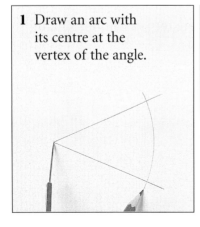	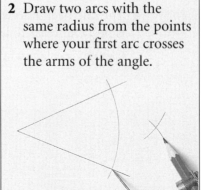	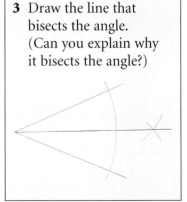

B1 Construct an angle of 45° by

  (a) constructing a perpendicular from a point to a line

  (b) bisecting the right angle

B2 Construct an angle of 30° by

  (a) constructing an equilateral triangle with compasses to get angles of 60°

  (b) bisecting one of the 60° angles

B3 Construct an angle of 75°.
Do not rub out the stages of your construction.

B4 Draw two lines that cross.

Bisect each of the four angles they make.

What do you notice about the angle bisectors?
Try to explain why this has happened.

B5 Draw an acute angle with arms at least 12 cm long.

Construct the bisector of the angle.

Mark a point P on the bisector well away from the vertex of the angle.

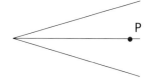

Construct a line from P that is perpendicular to the upper arm. Mark the point A where this perpendicular meets the upper arm.

Similarly, construct a perpendicular from P to a point B on the lower arm.

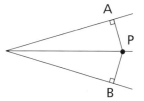

Draw a circle with its centre at P, going through A.
What do you notice about this circle?

Triangle investigation 2

Draw a large scalene triangle on A4 paper.

Draw the bisector of one angle of the triangle.
Now do the same for the other two angles.

What happens? Does it happen for other people's triangles?

Keep your drawing.

B6 In 'Triangle investigation 2' the three bisectors you drew should all have met at one point.

Using that point as the centre, draw a circle that just touches one side of the triangle.

What happens? Can you explain why?

B7 All the angles of this quadrilateral have been bisected. The bisectors have produced another quadrilateral in the middle of the original one.

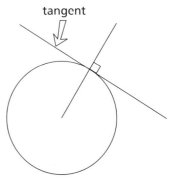

For each special quadrilateral below, if you bisect all its angles, do the bisectors produce another quadrilateral? If so, what sort?

(a) A rectangle that is not a square

(b) A parallelogram that is not a rhombus or rectangle

(c) A rhombus

(d) An isosceles trapezium (one with a line of reflection symmetry)

(e) A kite

Ⓒ *The perpendicular to a line from a point on the line*

P is a point on line *l*. Follow this construction to draw a line from P perpendicular to *l*.

1 Draw arcs with centre P and the same radius.	**2** Draw two arcs with equal radius from the points where your first arcs cross line *l*.	**3** Join P to the point where the last two arcs cross.

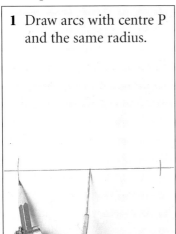
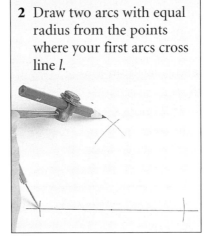
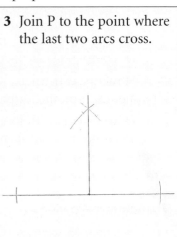

C1 A line that just touches a circle without crossing it is called a **tangent** to the circle.

Draw a circle with radius of about 5 cm.
Now follow these instructions to construct a tangent to your circle.

- Draw a radius of the circle.

- Construct the perpendicular to this radius at the point where it meets the circle.

tangent

C2 Draw a circle with radius about 5 cm.

From a point A on the circle, draw a line AB which crosses the circle at C.

Construct the perpendicular from C to point D on the circle.

Join AD with a straight line.
What do you notice about this line?
Did the same thing happen with other people's lines?

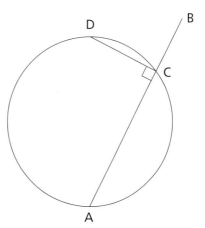

Ⓓ *The perpendicular bisector of a line segment*

In mathematics, 'line' nearly always means straight line.
A line goes on indefinitely in both directions.

a line

The proper name for a piece of a line is **line segment**.
But the word 'segment' can usually be left out without causing confusion.

a line segment

Carry out this construction.

Draw a horizontal line segment AB in the middle of an A4 sheet.	Use compasses to construct an isosceles triangle.	Construct more isosceles triangles above and below the line segment AB.
A ———————— B	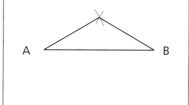	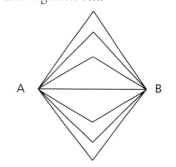

What do you notice about the vertices of the triangles?
Explain what you have found.

Every point on the dotted line is the same distance from A as it is from B. (We say each point is **equidistant** from A and B.)

The dotted line is perpendicular to line AB. It also bisects AB.

The symmetry of the construction on the previous page should help you see why these things are true.

The dotted line is called the **perpendicular bisector** of line AB.

Drawing the perpendicular bisector of a line segment

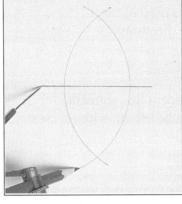

1 Draw a line segment. Draw an arc about this big with its centre at one end of the segment.

2 Keep your compasses the same radius. Draw an arc with its centre at the other end of the segment.

3 Draw a line through the points where the the two arcs cross. This is the perpendicular bisector of the original segment.

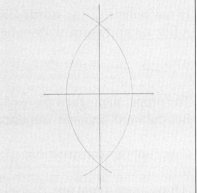

Triangle investigation 3

Draw a scalene triangle in the middle of a piece of paper.

Draw the perpendicular bisector of one side.

Now do the same for the other two sides.

What happens? Does it happen for other people's triangles?

Keep your drawing.

D1 In 'Triangle investigation 3' the three bisectors you drew should have all met at one point.

Using that point as the centre, draw a circle that goes through one the vertices of the triangle.

What happens? Can you explain why?

D2 If you are just given a circle (or part of one) you can use perpendicular bisectors to find its centre.
Follow these instructions.

Draw part of the way round a cup (or similar circular object).

Mark three points A, B and C anywhere on the circle.

Join A to B with a line segment. Join B to C with a line segment.

Construct the perpendicular bisector of each line segment.

Where the two bisectors intersect (cross one another) should be the centre of the circle.
See if you can draw the circle with compasses using this centre.

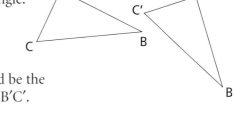

D3 Draw a triangle ABC.
Trace it, rotate the tracing paper and copy the triangle.
Letter the copy A′B′C′.

Join AA′. Draw the perpendicular bisector of AA′.
Join BB′. Draw the perpendicular bisector of BB′.

Where the perpendicular bisectors intersect should be the centre of the rotation that took triangle ABC to A′B′C′.
You can check this with your tracing paper.

As an accuracy check you can see if the perpendicular bisector of CC′ goes through this centre of rotation.

Test yourself with these questions

T1 Draw a line AB roughly 12 cm long.
Accurately construct a square with AB as one of its sides.
Use a ruler and compasses only,
but do not use the ruler for measuring.

T2 Using a ruler and compasses only, construct an angle of 150°.

T3 Draw a circle roughly 15 cm in diameter.
Mark four points A, B, C, D on it.
Join AB, AC, BD and CD.

Construct the bisector of angle ABD.
Do the same for angle ACD.

What do you notice about the two bisectors?

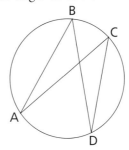

26 Gradient

You will

◆ calculate positive and negative gradients

◆ interpret gradients as rates

🄰 How steep?

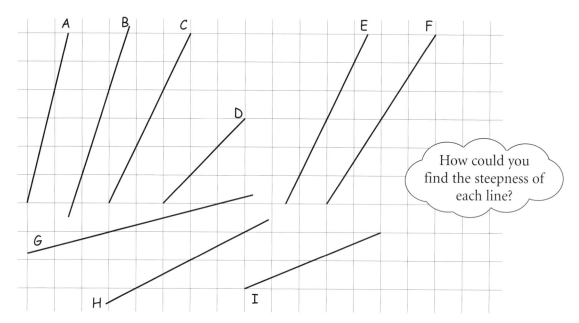

How could you find the steepness of each line?

A1 Find the gradient of each line in the diagram below.

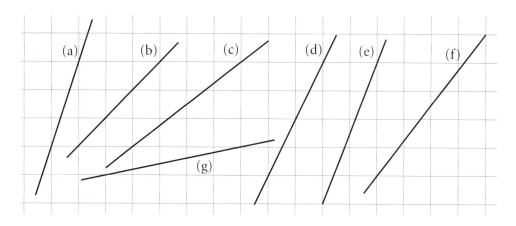

A2 Find the gradient of the line joining the points with coordinates (1,1) and (6,4).

A3 A county council uses the rule that the gradient of a wheelchair ramp must not be above 0.083.

Which of these ramps would be suitable for wheelchairs?

These sketches are not to scale.

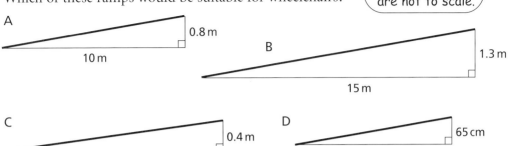

A4 Peter and Vicky are planning some walks.
They draw a sketch for each hill.

The gradient of each dotted line is the average gradient for each hill.

(a) Find the average gradient of each hill, correct to 3 decimal places.

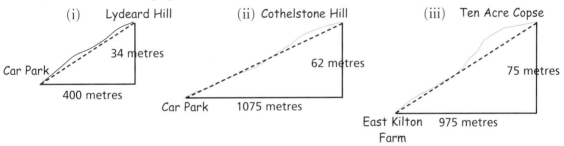

(b) Which hill has the highest average gradient?

A5 Stac Pollaidh is a mountain in Scotland.
The peak is 613 m above sea level.

Stephen and Carol are going to climb it starting at the car park at the foot of the mountain (90 m above sea level).

According to the map, the horizontal distance is 1050 m.

(a) What is the total height of their climb?

(b) What is the average gradient of the climb as a decimal?

A6 Scafell Pike is the highest mountain in England.
It is 977 m above sea level.

Becky and Ian are going to climb it starting at the car park at West Water (80 m above sea level).

According to the map, the horizontal distance is 3800 m.

What is the average gradient of the climb as a decimal?

OCR

Ⓑ *Gradient and rates*

It takes 8 minutes to fill a container of water.
Water flows in at a slow steady rate.

This graph shows the volume of water in
the container during these 8 minutes.

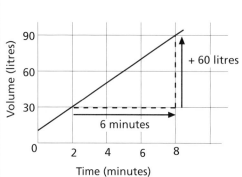

For a horizontal **increase** of 6 minutes
there is a vertical **increase** of 60 litres.

The **gradient** is $\dfrac{60}{6} = 10$

60 **litres** and
6 **minutes** so
10 **litres per minute**.

So the rate of flow of the water is
10 litres per minute.

B1 What rates of flow are shown by the following graphs?

(a)

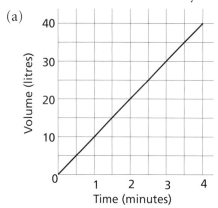

(b)

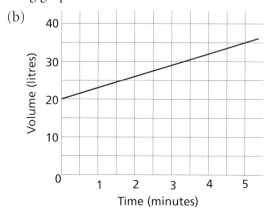

(c)

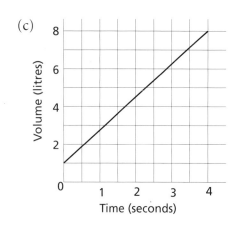

(d)

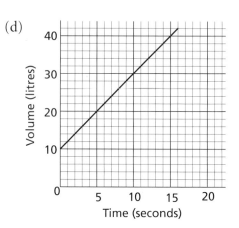

B2 This graph shows the distance Jamie cycled in the first 24 minutes of his journey.

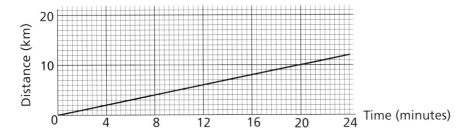

(a) Work out the gradient of this line.

(b) Which of these statements is true?

    A   The gradient represents the speed in km per hour.

    B   The gradient represents the speed in metres per minute.

    C   The gradient represents the speed in km per minute.

B3 (a) For the graphs below, work out the gradient of each line.

    (b) What does each gradient represent?

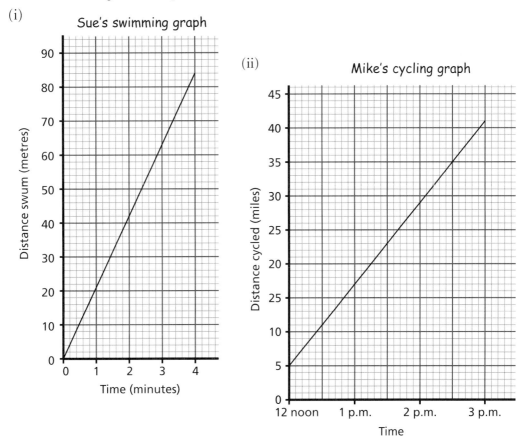

B4 (a) Work out the gradient of each line, correct to 1 d.p.

(b) What does each gradient represent?

(i)

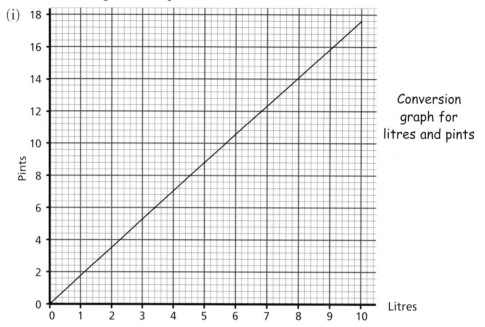

Conversion graph for litres and pints

(ii)

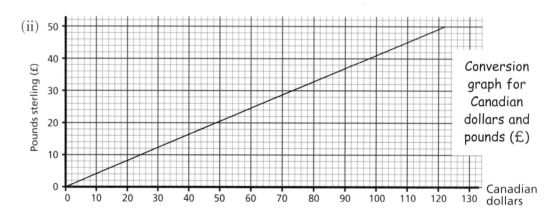

Conversion graph for Canadian dollars and pounds (£)

B5 The graph shows the speed of a marble rolling down a slope during a 5 second period.

(a) Work out the gradient of the line.

(b) What does the gradient of the line represent?

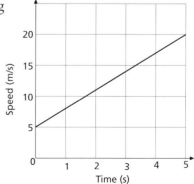

OCR

ℂ *Positive and negative gradients*

The gradient of a straight line is $\dfrac{\text{vertical change}}{\text{horizontal change}}$.

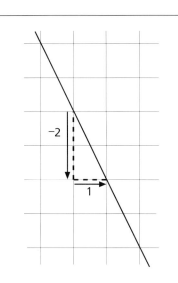

For a horizontal **increase** of 1 there is a vertical **decrease** of 2.

So **gradient** is $\dfrac{-2}{1} = -2$.

This sketch shows the volume of water in a tank.

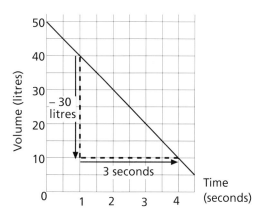

For a horizontal **increase** of 3 seconds there is a vertical **decrease** of 30 litres.

So **gradient** is $\dfrac{-30}{3} = -10$.

So the rate of flow is -10 litres per second.
The volume of water is **decreasing**.

C1 Find the gradient of each line in the diagram on the right.

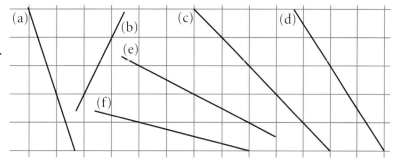

C2 This graph shows the volume of oil in a tank.

(a) Calculate the gradient of the line.

(b) What happens to the oil in the tank during these 5 minutes?

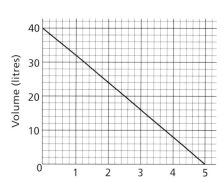

C3 This graph shows the volume of water in a tank.

(a) Find the gradient of each straight-line segment A, B, C and D.

(b) What do you think happened at

    (i) 5 minutes from the start

    (ii) 15 minutes from the start

    (iii) 25 minutes from the start

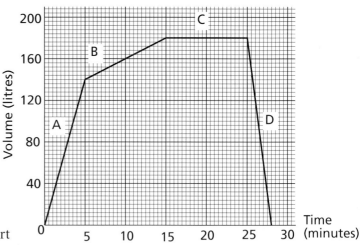

Test yourself with these questions

T1 (a) Find the gradient of this ramp (to 2 d.p.).

    (b) A fork-lift truck can safely climb a ramp with a gradient of less than 0.17 Can the fork-lift truck safely use this ramp?

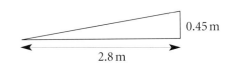

T2 Find the gradient of the line through the points A and B.

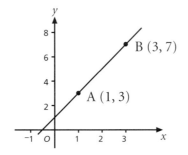

T3 This graph shows the volume of water in a container during 6 seconds.

    (a) Find the gradient of the line.

    (b) What does the gradient represent?

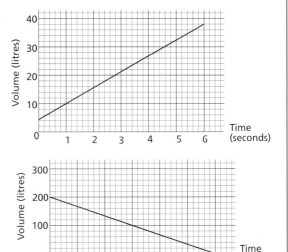

T4 This graph shows the volume of oil in a tank during 6 minutes.

    (a) Find the gradient of the line, correct to 1 d.p.

    (b) What is happening to the oil in the tank during these 6 minutes?

27 Indices

You should know how to work out the value of powers such as 4^3 (4 to the power 3).
This work will help you

◆ work with negative indices (and evaluate say 7^0)
◆ use the rules for multiplying and dividing powers of the same number
◆ use the rules for indices to simplify algebraic expressions

A Reptoids

Each 'active end' of a reptoid produces a small reptoid each day.

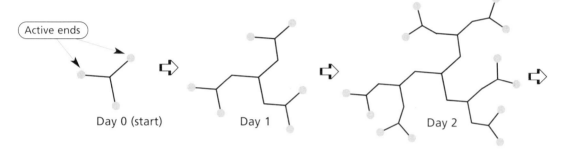

Active ends

Day 0 (start)   Day 1   Day 2

• How many active ends will this reptoid have on day 3, day 4, … ?
• What about day n?
• Investigate for other reptoids.

B Evaluating powers

B1 Evaluate these.

    (a) $2^4 + 5$ (b) 3×2^3 (c) $3^2 \times 3^0$ (d) $5^2 + 2^5$

    (e) 10×3^4 (f) $10^2 \div 5$ (g) $3^3 \div 9$ (h) $5^1 \times 4^2$

B2 Find the missing number in these statements.

    (a) $6^{\blacksquare} = 36$ (b) $\blacksquare^3 = 125$ (c) $7^{\blacksquare} = 7$ (d) $3^{\blacksquare} = 1$

B3 Work out the value of $2^n + 1$ when

    (a) $n = 3$ (b) $n = 5$ (c) $n = 0$ (d) $n = 1$

B4 For each of the following write down the value of n.

    (a) $3^n = 81$ (b) $6^n = 6$ (c) $n^3 = 64$ (d) $n^7 = 1$

    (e) $10^n = 100$ (f) $3^n = 243$ (g) $10^n = 100\,000$ (h) $n^3 = 125$

B5 Work out the missing number in

(a) $3^{\blacksquare} + 5 = 14$ (b) $\blacksquare^3 \div 4 = 16$ (c) $5^{\blacksquare} + 4 = 5$

(d) $\blacksquare^3 - 18 = 9$ (e) $2^{\blacksquare} \times 3 = 48$ (f) $8^{\blacksquare} \times 2 = 16$

B6 Work out the value of k in

(a) $2^k + 9 = 17$ (b) $6 \times 10^k = 600\,000$ (c) $k \times 5^2 = 150$

(d) $3^3 + 2^k = 29$ (e) $3^k \times 2 = 162$ (f) $5^3 - 4^k = 121$

B7 Work out the value of these expressions when $n = 3$.

(a) $n^2 \times n^2$ (b) $2n^2$ (c) $n^2 + n^2$

(d) $n^2 \times n$ (e) $n^2 + n^2 + n^2 + n^2$ (f) n^4

(g) $n^2 + n^1$ (h) n^3 (i) $4n^2$

$n^0 = 1$

For example $1^0 = 1,\ 2^0 = 1,\ 3^0 = 1, 4^0 = 1, \ldots$

C *Multiplying*

To **multiply** powers of the same number, **add** the indices.

Examples

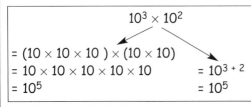

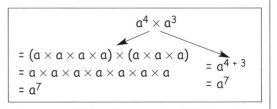

C1 Find four pairs of equivalent expressions.

A $3^5 \times 3^2$ **B** 3×3^7 **C** 3^7 **D** 3^{10}

E 3^9 **F** $3^3 \times 3^7$ **G** 3^8 **H** $3^5 \times 3^4$

C2 Copy and complete

(a) $4^3 \times 4^2 = 4^{\blacksquare}$ (b) $5^6 \times 5^3 = 5^{\blacksquare}$ (c) $3^4 \times 3^{\blacksquare} = 3^{11}$

(d) $8^2 \times 8^{\blacksquare} = 8^7$ (e) $7^{\blacksquare} \times 7^3 = 7^4$ (f) $9 \times 9^2 \times 9^5 = 9^{\blacksquare}$

C3 Write the answers to these using indices.

(a) $6^2 \times 6^7$ (b) $2^4 \times 2^9$ (c) $10^5 \times 10^7$ (d) $5^6 \times 5$

(e) $3^4 \times 3^2 \times 3^5$ (f) $8 \times 8^6 \times 8^2$ (g) $9^3 \times 9 \times 9^3$ (h) $2^4 \times 2^5 \times 2^3$

C4 This table shows some powers of 7.

7^2	7^3	7^4	7^5	7^6	7^7	7^8
49	343	2401	16 807	117 649	823 543	5 764 801

Use the results in the table to evaluate

(a) 49×343 (b) $343 \times 16\,807$ (c) 2401×343 (d) $823\,543 \times 7$

C5 In this wall, each expression is written as a power and is found by **multiplying** the two powers on the bricks below.

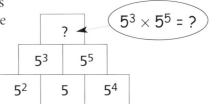

What should be on the top brick?

C6 Copy and complete these multiplication walls.

(a) 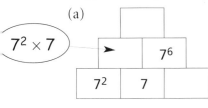   (b) 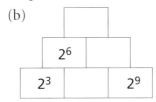   (c)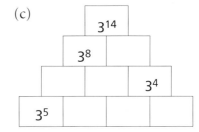

C7 Simplify each of these.

(a) $b \times b \times b \times b$ (b) $h^3 \times h^2$ (c) $a^4 \times a^5$ (d) $k \times k^5$

(e) $d^7 \times d$ (f) $m^2 \times m^4 \times m^6$ (g) $p \times p^9 \times p^3$ (h) $n \times n^7 \times n$

C8 Copy and complete

(a) $y^5 \times y^3 = y^\blacksquare$ (b) $n^2 \times n^\blacksquare = n^6$ (c) $h^\blacksquare \times h^2 \times h^5 = h^{11}$

(d) $k \times k^\blacksquare = k^6$ (e) $b^\blacksquare \times b^3 = b^3$ (f) $p^4 \times p^\blacksquare \times p^3 = p^{24}$

C9 Copy and complete these multiplication walls.

(a) 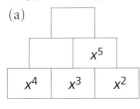   (b) 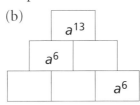   (c)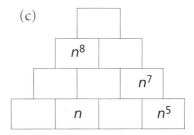

C10 Decide if each statement is true or false.

(a) $(2^3)^2 = 2^3 \times 2^3$ (b) $(2^3)^2 = 2^3 + 2^3$ (c) $(4^2)^3 = 4^2 + 4^2 + 4^2$

(d) $(4^2)^3 = 4^2 \times 4^2 \times 4^2$ (e) $(5^4)^2 = 5^4 \times 5^4$ (f) $(5^4)^2 = 5^4 + 5^4$

C11 Copy and complete these statements.

(a) $(7^3)^2 = 7^3 \times 7^3 = 7^\blacksquare$ (b) $(2^4)^3 = 2^4 \times 2^4 \times 2^4 = 2^\blacksquare$

C12 Simplify these.

(a) $(3^3)^2$ (b) $(2^3)^3$ (c) $(4^5)^2$ (d) $(3^4)^3$

C13 Copy and complete $(x^5)^2 = x^5 \times x^5 = x^\blacksquare$.

C14 Simplify these.

(a) $(p^4)^2$ (b) $(x^2)^2$ (c) $(k^0)^4$ (d) $(n^5)^3$

C15 Copy and complete these statements.

(a) $(5^2)^\blacksquare = 5^8$ (b) $(10^\blacksquare)^3 = 10^{12}$ (c) $(n^5)^\blacksquare = n^{10}$ (d) $(x^\blacksquare)^7 = x^{14}$

Ⓓ *Multiplying further*

These are multiplication walls.

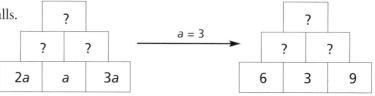

D1 (a) Find four pairs of matching expressions.

A $2n \times 6n$ **B** $4n^2 \times 3n$ **C** $8n^2$ **D** $12n^2$ **E** $2n \times 10n$

F $12n^3$ **G** $2n \times 4n$ **H** $2n^3 \times 4n^2$ **I** $8n^5$

(b) Which is the odd one out?

D2 Simplify these.

(a) $3p \times 5p$ (b) $n \times 2n$ (c) $2m^2 \times 9m$ (d) $7d^3 \times d^4$

(e) $2h^5 \times 3h^2$ (f) $4x^3 \times 5x^3$ (g) $3n^2 \times n^4 \times 7n^3$ (h) $2n^3 \times 2n \times 2n^5$

D3 Copy and complete these multiplication walls.

(a) 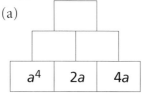 (b) (c)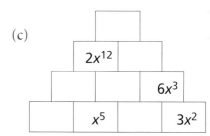

D4 Copy and complete

(a) $\blacksquare \times 4p = 8p^2$ (b) $7m^2 \times \blacksquare = 7m^3$ (c) $\blacksquare \times 4d = 12d^3$

(d) $3h^5 \times \blacksquare = 6h^{10}$ (e) $\blacksquare \times 3h^6 = 18h^7$ (f) $5p^4 \times \blacksquare = 15p^{12}$

D5 Solve the 'cover-up' puzzle on sheet P55.

D6 (a) Copy and complete $(2p)^3 = 2p \times 2p \times 2p = \blacksquare$

(b) Copy and complete $(3n^3)^2 = 3n^3 \times 3n^3 = \blacksquare$

D7 Simplify

(a) $(2y)^3$ (b) $(3n)^2$ (c) $(2p^5)^2$ (d) $(4m^3)^3$

Examples of multiplying expressions

$$\begin{aligned} 5a \times 3a &= 5 \times a \times 3 \times a \\ &= 5 \times 3 \times a \times a \\ &= 15a^2 \end{aligned}$$

$$\begin{aligned} 2p^2 \times 3p^5 &= 2 \times p^2 \times 3 \times p^5 \\ &= 2 \times 3 \times p^2 \times p^5 \\ &= 6p^7 \end{aligned}$$

$$\begin{aligned} (5q^3)^2 &= 5q^3 \times 5q^3 \\ &= 5 \times q^3 \times 5 \times q^3 \\ &= 5 \times 5 \times q^3 \times q^3 \\ &= 25q^6 \end{aligned}$$

E Dividing

$$\begin{aligned} 5^6 \div 5^4 &= \frac{5^6}{5^4} \\ &= \frac{5 \times 5 \times 5 \times 5 \times 5 \times 5}{5 \times 5 \times 5 \times 5} \\ &= \frac{5 \times 5 \times \cancel{5}^1 \times \cancel{5}^1 \times \cancel{5}^1 \times \cancel{5}^1}{\cancel{5}_1 \times \cancel{5}_1 \times \cancel{5}_1 \times \cancel{5}_1} \\ &= 5 \times 5 \\ &= 5^2 \end{aligned}$$

$$\begin{aligned} 3^7 \div 3^3 &= \frac{3^7}{3^3} \\ &= \frac{3 \times 3 \times 3 \times 3 \times 3 \times 3 \times 3}{3 \times 3 \times 3} \\ &= \frac{3 \times 3 \times 3 \times 3 \times \cancel{3} \times \cancel{3} \times \cancel{3}}{\cancel{3} \times \cancel{3} \times \cancel{3}} \\ &= 3 \times 3 \times 3 \times 3 \\ &= 3^4 \end{aligned}$$

We usually leave out 1s as multiplying or dividing by 1 doesn't affect a number.

- Can you see a rule for dividing powers?

E1 (a) Find four pairs of equivalent expressions.

A   **B**   **C**   **D**   **E** 2^4 **F**   **G** 2^2

H $2^{10} \div 2^5$ **I** 2

(b) Which is the odd one out?

E2 Write the answers to these using indices.

(a) $7^{10} \div 7^2$ (b) $2^9 \div 2^4$ (c) $10^7 \div 10^2$ (d) $5^6 \div 5$

(e) $\dfrac{3^9}{3^7}$ (f) $\dfrac{8^6}{8}$ (g) $\dfrac{10^{12}}{10^3}$ (h) $\dfrac{5^4}{5^4}$

E3 Find the value of n in each statement.

(a) $5^8 \div 5^2 = 5^n$ (b) $6^9 \div 6^n = 6^2$ (c) $9^n \div 9^8 = 9^3$ (d) $3^n \div 3^2 = 3^{12}$

(e) $\dfrac{9^{16}}{9^n} = 9^2$ (f) $\dfrac{2^n}{2} = 2^5$ (g) $\dfrac{7^7}{7^n} = 7^3$ (h) $\dfrac{3^n}{3^8} = 3^0$

E4 Write the answers to these using indices.

(a) $6^3 \times 6^5$ (b) $\dfrac{7^5}{7^3}$ (c) $\dfrac{2^5 \times 2^4}{2^3}$ (d) $\dfrac{3^5 \times 3^6}{3^9}$

(e) $\dfrac{(5^4)^2}{5^5}$ (f) $\dfrac{2^8}{2 \times 2^4}$ (g) $\dfrac{7 \times 7^6}{7^2 \times 7^3}$ (h) $\dfrac{8^3 \times 8^4}{8^2 \times 8^3}$

E5 Work out $\dfrac{2^6 \times 2^2}{2^3 \times 2^5}$.

E6 This table shows some powers of 8.

8^2	8^3	8^4	8^5	8^6	8^7	8^8
64	512	4096	32 768	262 144	2 097 152	16 777 216

Use the results in the table to evaluate

(a) $\dfrac{32\,768}{512}$ (b) $\dfrac{262\,144}{512}$ (c) $\dfrac{16\,777\,216}{2\,097\,152}$ (d) $\dfrac{64 \times 262\,144}{16\,777\,216}$

E7 Find the missing number in each statement.

(a) $\dfrac{p^5}{p^3} = p^{\blacksquare}$ (b) $\dfrac{x^7}{x^2} = x^{\blacksquare}$ (c) $\dfrac{n^{12}}{n^9} = n^{\blacksquare}$ (d) $\dfrac{a^{10}}{a} = a^{\blacksquare}$

E8 Write the answers to these using indices.

(a) $\dfrac{h^7}{h^3}$ (b) $\dfrac{n^9}{n^4}$ (c) $\dfrac{x^4}{x}$ (d) $\dfrac{d^{12}}{d^3}$ (e) $\dfrac{a^5}{a^4}$

E9 Find the missing number in each statement.

(a) $\dfrac{b^7}{b^{\blacksquare}} = b^5$ (b) $\dfrac{k^{10}}{k^{\blacksquare}} = k^2$ (c) $\dfrac{m^{\blacksquare}}{m} = m^3$ (d) $\dfrac{h^{\blacksquare}}{h^6} = h$

E10 Work out the value of each expression when $n = 3$.

(a) $\dfrac{n^4}{n^3}$ (b) $\dfrac{n^8}{n^6}$ (c) $\dfrac{n^{10}}{n^9}$ (d) $\dfrac{n^{14}}{n^{12}}$ (e) $\dfrac{n^6}{n^3}$

E11 Simplify

(a) $g^3 \times g^5$ (b) $\dfrac{w^6}{w^2}$ (c) $\dfrac{p^5 \times p}{p^2}$ (d) $\dfrac{h^5 \times h^6}{h^{10}}$

(e) $\dfrac{(y^3)^2}{y}$ (f) $\dfrac{h^8}{h \times h^3}$ (g) $\dfrac{q \times q^9}{q^3 \times q^4}$ (h) $\dfrac{z^5 \times z^3}{z^4 \times z^4}$

To **divide** powers of the same number, you can **subtract** the indices.

Examples

$$\frac{10^9}{10^7} = 10^{9-7}$$
$$= 10^2$$

$$\frac{8^6}{8^5} = 8^{6-5}$$
$$= 8^1$$
$$= 8$$

$$\frac{m^{13}}{m^9} = m^{13-9}$$
$$= m^4$$

F *Dividing further*

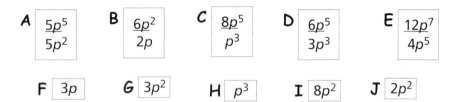

A $\dfrac{5p^5}{5p^2}$ **B** $\dfrac{6p^2}{2p}$ **C** $\dfrac{8p^5}{p^3}$ **D** $\dfrac{6p^5}{3p^3}$ **E** $\dfrac{12p^7}{4p^5}$

F $3p$ **G** $3p^2$ **H** p^3 **I** $8p^2$ **J** $2p^2$

- Can you find five pairs of equivalent expressions?

F1 Simplify these.

   (a) $\dfrac{5p^3}{p^2}$ (b) $\dfrac{8a^6}{4a^2}$ (c) $\dfrac{12y^4}{3y^3}$ (d) $\dfrac{8m^9}{8m^7}$ (e) $\dfrac{16n^7}{4n}$

F2 What are the missing numbers from each statement?

   (a) $\dfrac{3p^4}{3p^{\blacksquare}} = p$ (b) $\dfrac{10x^9}{\blacksquare x^3} = 5x^6$ (c) $\dfrac{14n^{\blacksquare}}{7n^2} = 2n^3$ (d) $\dfrac{\blacksquare k^7}{3k^{\blacksquare}} = 9k$

F3 Copy and complete each grid.
   Each space should contain an expression, × or ÷.

   Each row (from left to right) and each column (from top to bottom)
   should show a true statement.

(a)

	×	$6a^6$	=	
×	■	÷	■	
$2a^5$	×		=	$6a^6$
=	■	=	■	=
	÷		=	$5a^3$

(b)

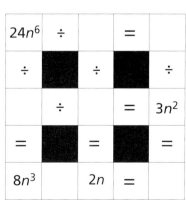

$24n^6$	÷		=	
÷	■	÷	■	÷
		÷	=	$3n^2$
=	■	=	■	=
$8n^3$		$2n$	=	

$\mathbb{G}$ *Even further*

One way to simplify divisions is to **cancel** common factors.

Examples

$$\dfrac{3^2}{3^4} = \dfrac{3 \times 3}{3 \times 3 \times 3 \times 3}$$

$$= \dfrac{1 \;\cancel{3} \times \cancel{3}}{{}_1\cancel{3} \times \cancel{3} \times 3 \times 3}$$

$$= \dfrac{1}{3 \times 3}$$

$$= \dfrac{1}{3^2}$$

$$\dfrac{10p^2}{5p^5} = \dfrac{10 \times p \times p}{5 \times p \times p \times p \times p \times p}$$

$$= \dfrac{{}^2\cancel{10} \times \cancel{p} \times \cancel{p}}{\cancel{5} \times \cancel{p} \times \cancel{p} \times p \times p \times p}$$

$$= \dfrac{2}{p \times p \times p}$$

$$= \dfrac{2}{p^3}$$

$$\dfrac{6a^3}{10a^4} = \dfrac{6 \times a \times a \times a}{10 \times a \times a \times a \times a}$$

$$= \dfrac{{}^3\cancel{6} \times \cancel{a} \times \cancel{a} \times \cancel{a}}{{}^5\cancel{10} \times \cancel{a} \times \cancel{a} \times \cancel{a} \times a}$$

$$= \dfrac{3}{5 \times a}$$

$$= \dfrac{3}{5a}$$

G1 Simplify by cancelling

(a) $\dfrac{5^2}{5^6}$ (b) $\dfrac{7^3}{7^8}$ (c) $\dfrac{2^9}{2^{10}}$ (d) $\dfrac{5^6}{5^{12}}$ (e) $\dfrac{3}{3^5}$

(f) $\dfrac{p^2}{p^9}$ (g) $\dfrac{k^3}{k^7}$ (h) $\dfrac{n^3}{n^{12}}$ (i) $\dfrac{x^9}{x^{10}}$ (j) $\dfrac{y}{y^8}$

G2 Simplify by cancelling

(a) $\dfrac{3p^4}{p^7}$ (b) $\dfrac{10b^2}{5b^6}$ (c) $\dfrac{9x^3}{3x^8}$ (d) $\dfrac{m^8}{5m^5}$ (e) $\dfrac{a^{10}}{7a^9}$

(f) $\dfrac{12y^3}{24y^2}$ (g) $\dfrac{3n^7}{15n^5}$ (h) $\dfrac{30b^4}{3b^7}$ (i) $\dfrac{3x^7}{21x^{11}}$ (j) $\dfrac{6x^6}{18x^7}$

G3 Simplify by cancelling

(a) $\dfrac{8a^4}{6a^5}$ (b) $\dfrac{6n^7}{15n^5}$ (c) $\dfrac{12p^2}{28p^3}$ (d) $\dfrac{4x}{18x^3}$ (e) $\dfrac{15k^7}{25k^{10}}$

$\mathbb{H}$ *Negative indices*

A 'doublebug' doubles in length each day.

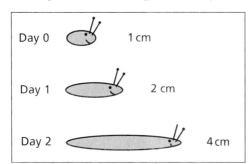

Day 0   1 cm

Day 1   2 cm

Day 2   4 cm

- How long will it be on day 3?
- How long was it on day $^-1$?
- What about day $^-2$, day $^-3$ … ?

H1 (a) Find three matching pairs.

A 3^{-4} B 3^{-2} C 2^{-3} D 4^{-3} E $\frac{1}{4^3}$ F $\frac{1}{2^3}$ G $\frac{1}{3^4}$

(b) Which is the odd one out?

H2 (a) Find four matching pairs.

A 3^{-2} B 2^{-3} C 4^{-4} D 6^{-1} E 4^{-2} F $\frac{1}{8}$ G $\frac{1}{6}$ H $\frac{1}{16}$ I $\frac{1}{9}$

(b) Which is the odd one out?

H3 2^{-4} is equivalent to the fraction $\frac{1}{16}$.

Write these as fractions.

(a) 3^{-3} (b) 7^{-1} (c) 9^{-2} (d) 4^{-3} (e) 11^{-1}

H4 (a) Find four matching pairs.

A $2^{-1} + 2^{-2}$ B $5^0 - 5^{-1}$ E $\frac{3}{4}$ G $\frac{-1}{5}$ H 1

C $8^{-1} \times 8^0$ D 4×2^{-2} F $\frac{1}{8}$ I $\frac{4}{5}$

(b) Which is the odd one out?

H5 $2^{-2} = \frac{1}{2^2} = \frac{1}{4} = 0.25$ as a decimal.

Work out the decimal value of

(a) 2^{-1} (b) 4^{-1} (c) 2^{-2} (d) 10^{-2} (e) 5^{-1}

H6 Write $5^{-1} + 5^0$ as a decimal.

H7 Write $10^{-1} + 10^{-3}$ as a decimal.

H8 a^{-2} is equivalent to the fraction $\frac{1}{a^2}$ in fractional form.

Write these in fractional form.

(a) x^{-3} (b) g^{-1} (c) n^{-2} (d) k^{-3} (e) p^{-1}

H9 Find the missing number in each statement below.

(a) $12^{\blacksquare} = \frac{1}{12}$ (b) $10^{\blacksquare} = 1$ (c) $2^{\blacksquare} = \frac{1}{4}$ (d) $4^{\blacksquare} = 0.25$

(e) $10^{\blacksquare} = 0.01$ (f) $5^{\blacksquare} = \frac{1}{25}$ (g) $\blacksquare^{-2} = \frac{1}{49}$ (h) $\blacksquare^{-3} = \frac{1}{125}$

$a^{-m} = \frac{1}{a^m}$

For example $2^{-4} = \frac{1}{2^4}$ $3^{-5} = \frac{1}{3^5}$ $5^{-1} = \frac{1}{5}$

More multiplying and dividing

A $3^5 \times 3^{-2} = ?$ **B** $2^{-5} \times 2^3 = ?$ **C** $2^{-3} \times 2^3 = ?$ **D** $2^{-1} \times 2^{-2} = ?$

E $10^2 \div 10^4 = ?$ **F** $\dfrac{3^3}{3^5} = ?$ **G** $\dfrac{7^2}{7^2} = ?$ **H** $\dfrac{2^3}{2^4} = ?$

I1 Write the answers to these as a single power.

(a) $3^4 \times 3^{-3}$ (b) $10^{-2} \times 10^5$ (c) $8^{-4} \times 8^4$ (d) $3^{-5} \times 3^3$

(e) $2^2 \times 2^{-7}$ (e) 9×9^{-2} (e) $2^{-4} \times 2^{-1}$ (e) $7^{-2} \times 7^9 \times 7^{-4}$

I2 Write the answers to these as a single power.

(a) $3^2 \div 3^4$ (a) $5^3 \div 5^6$ (a) $2^4 \div 2^5$ (a) $9 \div 9^6$

(e) $\dfrac{4^5}{4^7}$ (f) $\dfrac{7^3}{7^9}$ (g) $\dfrac{6^7}{6^8}$ (h) $\dfrac{10}{10^7}$

I3 Write the answers to these as a single power.

(a) $5^4 \times 5^{-2}$ (b) $2^3 \div 2^8$ (c) $7^{-9} \times 7^5$ (d) $6 \div 6^8$

(e) $7^{-3} \times 7^{-2}$ (f) $\dfrac{4^3}{4^7}$ (g) $2^9 \times 2^{-9}$ (h) $\dfrac{5^2}{5^3}$

I4 Simplify each of these.

(a) $p^4 \times p^{-5}$ (b) $q^3 \times q^{-3}$ (c) $r^{-7} \times r^9$ (d) $s^{-3} \times s$

(e) $\dfrac{w^2}{w^7}$ (f) $\dfrac{x^3}{x^6}$ (g) $\dfrac{y^8}{y^8}$ (h) $\dfrac{z}{z^5}$

I5 Simplify each of these.

(a) 11×11^{-7} (b) $2^6 \div 2^3$ (c) $a^{-4} \times a^3$ (d) $6 \div 6^4$

(e) $13^{-2} \times 13^{-5}$ (f) $\dfrac{b^8}{b^6}$ (g) $3^4 \times 3^{-4}$ (h) $\dfrac{2^3}{2^9}$

J Mixed questions

J1 Solve the equation $2^x = 64$.

J2 Work out the value of 5^n when $n = 0$.

J3 Write $2^4 \times 2 \times 2^8$ as a single power of 2.

J4 Simplify these.

(a) $p^2 \times p \times p^5$ (b) $5x \times 2x^3$ (c) $6a^3 \times 3a^2$ (d) $(2k)^4$

J5 Write as a fraction

(a) 2^{-3} (b) 4^{-1} (c) 3^{-2} (d) 10^{-5}

J6 Write these as single powers of 3.

(a) $3^5 \div 3^2$ (b) $\dfrac{3^8}{3^4}$ (c) $\dfrac{3^7}{3^{10}}$ (d) $\dfrac{3^9}{3^8}$

J7 Simplify these.

(a) $\dfrac{15x^5}{5x^3}$ (b) $\dfrac{4a^6}{8a^3}$ (c) $\dfrac{10b^2}{2b^7}$ (d) $\dfrac{4c^7}{12c^9}$

J8 Simplify these.

(a) $p^{-3} \times p^{-2}$ (b) $p^{-1} \times p^8$ (c) $p^5 \times p^{-9}$ (d) $p^5 \div p^7$

J9 Simplify these.

(a) $4n^{-1} \times 2n^{-2}$ (b) $\dfrac{x^5}{x^{-2}}$ (c) $\dfrac{5p^2}{p^6}$ (d) $\dfrac{12a}{6a^3}$

Test yourself with these questions

T1 For each of the following equations, write down the value of n.

(a) $2^n = 32$ (b) $n^3 = 125$ (c) $8^n = 8$

AQA(NEAB)1998

T2 Write the answers to these using indices.

(a) $5^3 \times 5^2$ (b) $\dfrac{4^8}{4^5}$ (c) $\dfrac{2^3 \times 2^4}{2^5}$

T3 Simplify these.

(a) $h^3 \times h^2$ (b) $\dfrac{m^7}{m^5}$ (c) $\dfrac{n^5 \times n^6}{n}$

T4 Evaluate

(a) 3^3 (b) 2^{-3} (c) 7^0

T5 Write these as single powers.

(a) $7^9 \times 7^{-2}$ (b) $\dfrac{6^3}{6^9}$ (c) $\dfrac{3 \times 3^4}{3^{10}}$

T6 Find the value of p when $2^p \times 5 = 40$.

T7 Simplify these.

(a) $h \times 8h$ (b) $3b^2 \times 2b^6$ (c) $k^3 \times k^{-1}$

T8 Simplify these.

(a) $(2x^4)^3$ (b) $\dfrac{12m^6}{3m}$ (c) $\dfrac{8h^2}{4h^6}$

28 Pythagoras

You will revise square roots

You will learn how to

◆ find the length of one side of a right angled triangle
if you know the lengths of the other two sides

◆ solve problems involving the lengths of sides of right angled triangles

A Areas of tilted squares

To get the area of a tilted square...

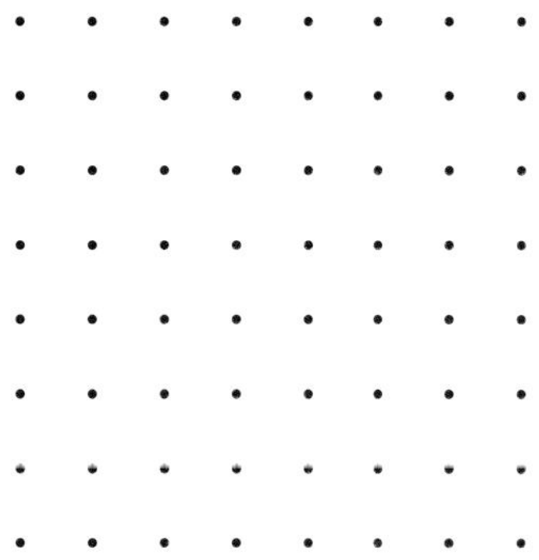

...find the area of a surrounding square, then subtract the area of the triangles...

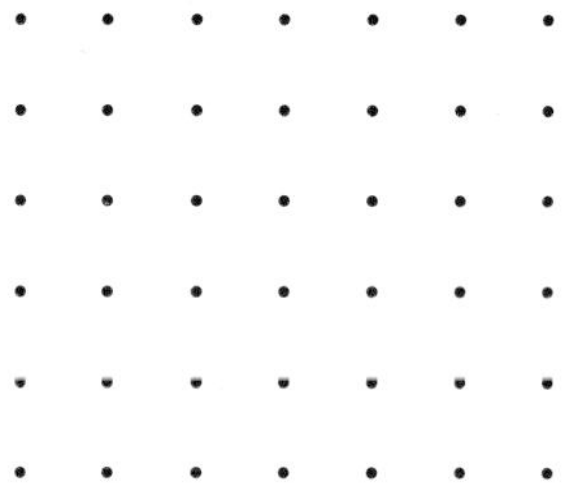

... **or** divide the square up like this and find the area of the parts.

A1 This is one side of a tilted square.

(a) Copy the line on to square dotty paper and complete the square.

(b) Work out the area of the square.

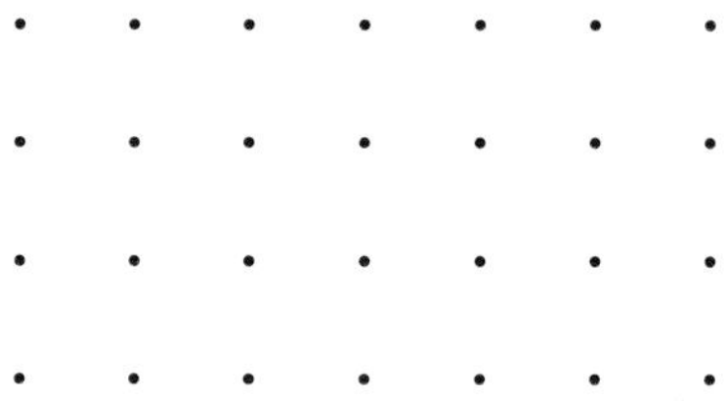

A2 Each of these is the side of a tilted square.
Draw each square and work out its area.

(a)   (b)    (c)

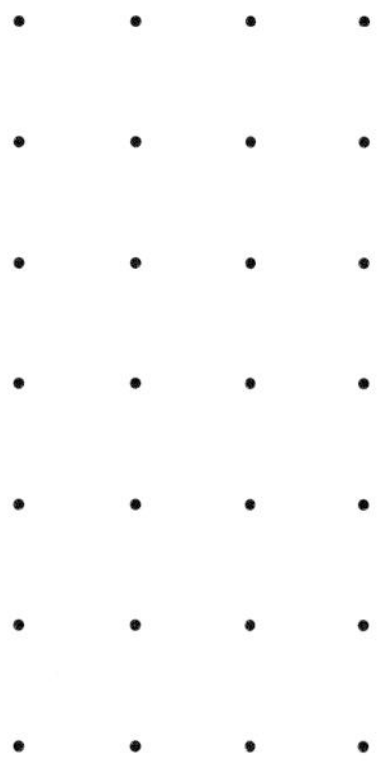

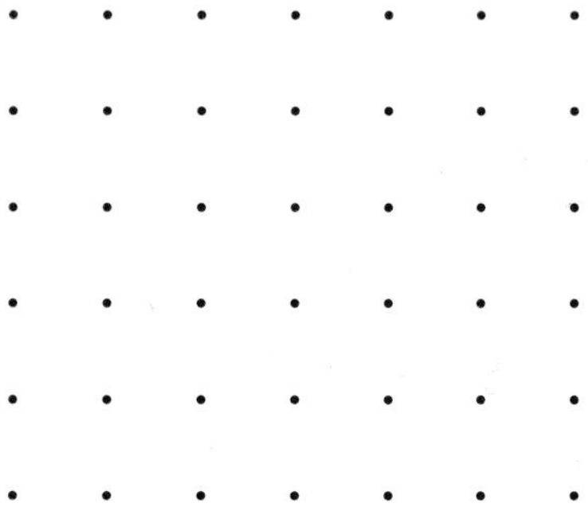

B Squares on right-angled triangles

The three squares Q, R and S are drawn on
the sides of a right angled triangle

Copy the drawing on to dotty paper.
Find and record the area of each square.

Repeat this process for different right-angled triangles.
Square S must be on the side opposite the right angle.
Record your results in a table.

Area of square Q	Area of square R	Area of square S

What happens?

B1 Find the missing areas of the squares on
these right-angled triangles.

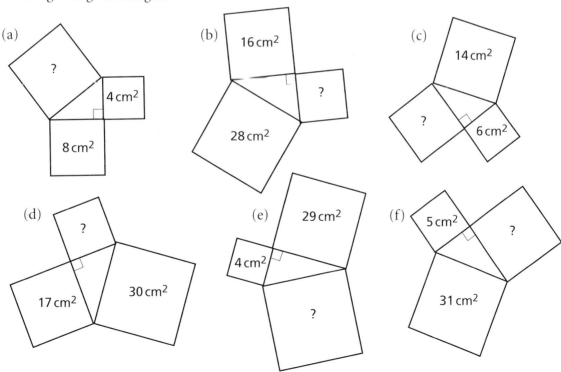

(a) ? , 4 cm², 8 cm²

(b) 16 cm², ?, 28 cm²

(c) 14 cm², ?, 6 cm²

(d) ?, 17 cm², 30 cm²

(e) 29 cm², 4 cm², ?

(f) 5 cm², ?, 31 cm²

Pythagoras's theorem

In a right-angled triangle the side opposite the right angle is called the **hypotenuse**.

You have found that the area of the square on the hypotenuse equals the total of the areas of the squares on the other two sides.

Here, Area C = Area A + Area B

This is known as Pythagoras's theorem.
Pythagoras was a Greek mathematician and mystic.
A theorem is a statement that can be proved true.

C

A

B

Using Pythagoras's theorem you can work with the lengths of sides as well as the areas of squares on them.

B2 (a) What is the area of the
  square drawn on side XY?

(b) What is the length of side XY?

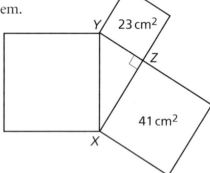

B3 What is the area of
the square drawn here? ⟹

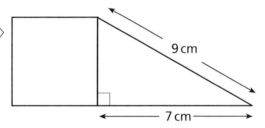

B4 Work out the missing area or length in each of these.

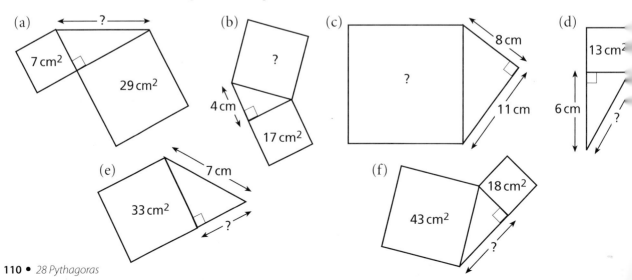

B5 Work out the missing area or length in each of these.

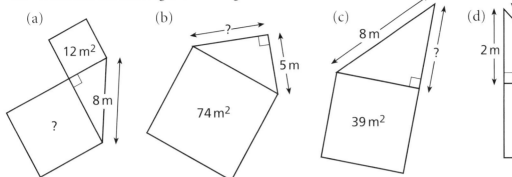

(a)

12 m²

8 m

?

(b)

?

5 m

74 m²

(c)

8 m

?

39 m²

(d)

2 m

9 m

?

Pythagoras in practice

Pythagoras is useful for working out lengths when designing and constructing things.

You don't have to draw squares on the sides of the right-angled triangle you are using.
You can think of Pythagoras just in terms of the lengths of the sides, as shown here.

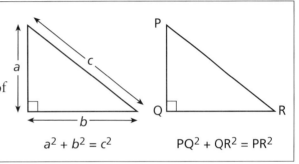

a

c

b

$a^2 + b^2 = c^2$

P

Q

R

$PQ^2 + QR^2 = PR^2$

B6 (a) Use Pythagoras to find out what length side LN should be.

(b) Now draw the triangle accurately with a ruler and set square.
Measure the length of LN and see if it agrees with the length you calculated.

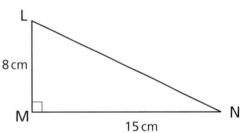

L

8 cm

M

15 cm

N

B7 Work out the missing lengths here.

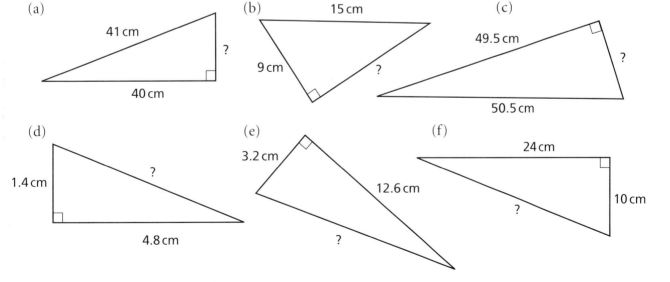

(a)

41 cm

?

40 cm

(b)

15 cm

9 cm

?

(c)

49.5 cm

?

50.5 cm

(d)

1.4 cm

?

4.8 cm

(e)

3.2 cm

12.6 cm

?

(f)

24 cm

10 cm

?

B8 People marking out sports pitches need to mark lines at right angles. They sometimes use a rope divided into 12 equal spaces to form a 3, 4, 5 triangle.

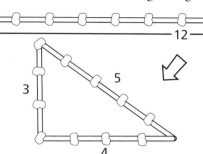

Use Pythagoras to check that this produces a right angle. Show your working.

C *Square roots - a reminder*

If a square has an area of 16 cm², you know the length of its side is 4 cm because 4^2 is 16.

Remember that 4 is the **square root** of 16.

C1 Copy and complete this table.

Number	Square root
1	
4	
	3
	4
25	
36	
	7

C2 What is the square root of each of these numbers?
(a) 81   (b) 100   (c) 121   (d) 400

We use the symbol $\sqrt{}$ to mean square root.
So $\sqrt{16}$ means the square root of 16.

C3 Work out the value of these.
(a) $\sqrt{49}$ (b) $\sqrt{9}$ (c) $\sqrt{144}$ (d) $\sqrt{169}$

C4 For each of these square roots
(i) first write down a rough answer.
(ii) find the value to 2 decimal places on a calculator.
(a) $\sqrt{10}$ (b) $\sqrt{2.5}$ (c) $\sqrt{150}$ (d) $\sqrt{15}$
(e) $\sqrt{200}$ (f) $\sqrt{20}$ (g) $\sqrt{42}$ (h) $\sqrt{420}$
(i) $\sqrt{85}$ (j) $\sqrt{8.5}$ (k) $\sqrt{805}$ (l) $\sqrt{50}$
(m) $\sqrt{500}$ (n) $\sqrt{5}$ (o) $\sqrt{0.5}$ (p) $\sqrt{5000}$

C5 Use a square root button to work out the missing lengths here.
Give your answers to 1 decimal place.

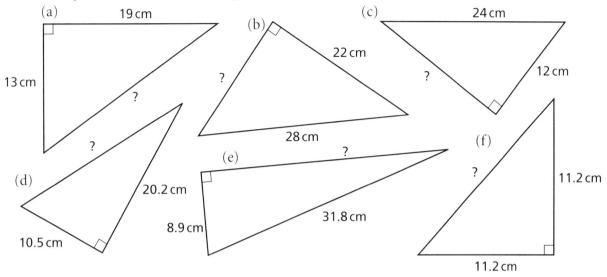

(a) 19 cm, 13 cm, ?

(b) 22 cm, 28 cm, ?

(c) 24 cm, 12 cm, ?

(d) 20.2 cm, 10.5 cm, ?

(e) 8.9 cm, 31.8 cm, ?

(f) 11.2 cm, 11.2 cm, ?

C6 (a) Use Pythagoras to work out side PQ to the nearest 0.1 cm.

(b) Now draw the triangle accurately with a ruler and set square.
Measure the length of PQ and see if it agrees with the length you calculated.

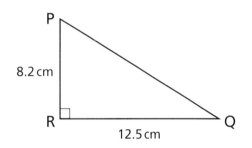

8.2 cm, 12.5 cm

Ⓓ *Using Pythagoras*

D1 A certain exercise book is 14.0 cm wide by 20.0 cm high.

(a) How long is the longest straight line you can draw on a single page of the book?

(b) How long is the longest straight line you can draw on a double page?

D2 Measure the height and width of your own exercise book.
Repeat the calculations in D1 for your own book.
Measure to check your answers.

D3 This is the plan of a rectangular field.
There is a footpath across the field from A to C.

How much shorter is it to use the footpath than to walk from A to B and then to C?

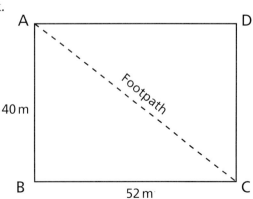

40 m, 52 m

D4 Points A and B are plotted on a grid on centimetre squared paper.

   (a) How far is it in a straight line from A to B?

   (b) How long would a straight line from (2, 2) to (14, 7) be? (Draw them on a grid or make a sketch if you need to.)

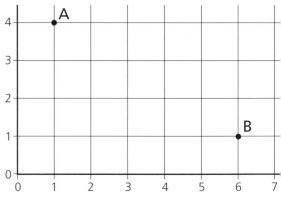

D5 (a) Calculate the lengths of the sides of this quadrilateral.

   (b) Use your working to say whether it is exactly a rhombus.

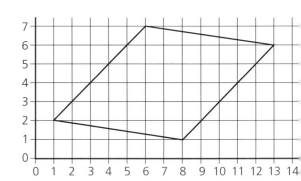

D6 A helicopter flies 26 km north from a heliport, then 19 km west. How far is it from the heliport now?

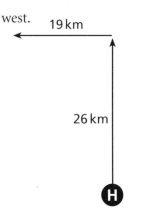

D7 How long is a straight line joining each pair of points if they are plotted on a centimetre squared grid? Give your answers to 1 decimal place.

   (a) (1, 3) to (5, 7)   (b) (2, 4) to (8, 1)   (c) (5, 0) to (7, −3)

   (d) (−2, 3) to (−4, 1)   (e) (−2, 6) to (2, 8)   (f) (3, 1) to (−6, 4)

   (g) (6, 3) to ((2, −4)   (h) (11, −1) to (8, 3)   (i) (−7, 5) to (−4, −5)

D8 A bird flies 8 km west from a lighthouse. It then flies south.

   How far south has it flown when it is 22 km from the lighthouse?

Test yourself with these questions

T1 Find the missing areas.

(a)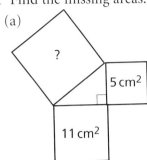

(b)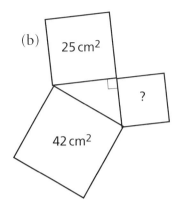

(c)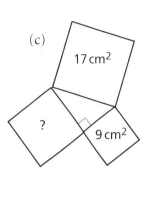

T2 Find the missing lengths (to 1 d.p.)

(a)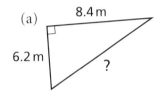

(b)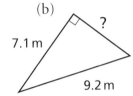

(c)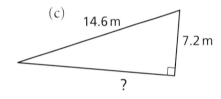

T3 A lifeboat travels 24 km east from its station and then 11km south.
It then travels in a straight line back to its station.
How far has it gone altogether?

T4 A ladder 3.0 m long rests against a vertical wall.
The foot of the ladder is 0.9 m away from the bottom of the wall,
on horizontal ground.

(a) Draw a sketch and label it with these measurements.

(b) Calculate the distance from the top of the ladder to the bottom of the wall.

T5 This trapezium has two right angles.
Calculate (to 1 d.p.)

(a) the length of the fourth side

(b) the length of each of its diagonals

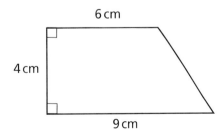

29 Looking at expressions

You should know how to
- multiply out brackets such as $4(2x - 3)$
- simplify expressions such as $4n^3 \times 2n$ and $4n^3 \div 2n$

This work will help you
- gather like terms in expressions involving powers
- multiply and divide expressions such as $4ab^3$ and $2ab$
- multiply out complex brackets such as $4x(xy + 3x)$
- factorise expressions such as $6x^2y - 9xy^2$
- form and simplify formulas

A Gathering like terms

A	$3^2 + 3^2$	B	6×3^2	C	3^4	D	$8 \times 3^2 - 2 \times 3^2$

E	$2 \times 3^2 + 5 \times 3^2$	F	$3^2 \times 3^2$	G	7×3^2	H	2×3^2

- Can you find four pairs of equivalent expressions?

A	$x^2 + x^2$	B	$3x^4 - 2x^4$	C	$3x^3 + x$	D	$2x^2 - x$

E	$x^2 + 2x + x^2 - 3x$	F	$2x^3 + 2x + x^3 - x$	G	x^4	H	$2x^2$

- Can you find four pairs of equivalent expressions?

A1 Simplify the following by collecting like terms.
- (a) $8 + 3n - 7 + 2n$
- (b) $7p + 6 - 5p - 9$
- (c) $5k - 7 - 9k + 12$

A2 Simplify the following by collecting like terms.
- (a) $n^2 + 3n + n^2 + 2n$
- (b) $3a^2 + 5a - a^2$
- (c) $k^2 - k + 2k^2 + 6k$
- (d) $7m + 4m^2 - 5m + 2$
- (e) $g^2 + g^3 + 4g^2 + g^3$
- (f) $5h^2 + 2h - 3h^2 - 5h$
- (g) $2x + 5x^2 - 3x + 2x^2$
- (h) $y^2 + y + y^2 - 7y + 3$

A3 Find the value of each expression when $n = 5$.
- (a) $n^2 + n^2 + n^2 + n^2 - 2$
- (b) $n^3 + 8n - n^3$
- (c) $8n^2 - 7n^2 + n + 1$
- (d) $2n^2 + 3n - n^2 - 4n$

A4

A $a - a^2$ **B** $a^2 + 2a$ **C** $3a^2 - a + 6$

D $a^2 + 2a - 1$ **E** $3 - 2a$

(a) Find pairs of the above expressions that add to give

    (i) $2a^2 + 4a - 1$ (ii) $a^2 + 3$ (iii) $2a^2 + 6$

    (iv) $4a^2 + a + 5$ (v) $3a - 1$ (vi) $a^2 + 2$

(b) Find three of the above expressions that add to give $a + 2$.

B *Multiplying*

Examples

$p(p - 3) = p \times p - p \times 3$	$2n(5 + n^2) = 2n \times 5 + 2n \times n^2$	$3b(2b - 5) = 3b \times 2b - 3b \times 5$
$\quad\quad = p^2 - 3p$	$\quad\quad\quad = 10n + 2n^3$	$\quad\quad\quad = 6b^2 - 15b$

B1 Multiply out the brackets from

    (a) $n(n + 7)$ (b) $m(3 + m)$ (c) $3(2a - 5)$ (d) $h(h - 9)$

    (e) $k(10 - k)$ (f) $2w(w + 7)$ (g) $3x(x - 6)$ (h) $6n(2 - n)$

B2 Find the missing expressions in these statements

    (a) $d(\blacksquare) = d^2 + 5d$ (b) $2n(\blacksquare) = 2n^2 - 8n$

    (c) $3p(\blacksquare) = 15p - 3p^2$ (d) $5k(\blacksquare) = 20k + 5k^2$

B3 Find five pairs of matching expressions.

A $b(b^2 + 5)$ **B** $2b(5b + 6)$ **C** $2(5b^2 + 3)$ **D** $3b(2 + 5b)$ **E** $2b(3b + 5)$

F $10b^2 + 6$ **G** $6b^2 + 10b$ **H** $10b^2 + 12b$ **I** $6b + 15b^2$ **J** $b^3 + 5b$

B4 Multiply out the brackets from

    (a) $2h(7h - 5)$ (b) $3a(a^2 - 4)$ (c) $n^2(n^3 - 5)$ (d) $3k(4k + 5)$

    (e) $5d(3 - 2d^2)$ (f) $7p(3 - 4p)$ (g) $2b^2(3b + 1)$ (h) $3w^2(1 + 2w^2)$

B5 3 x $2x$ $5x$ x^2 $2x + 5$ $x + 1$

Find pairs of the above expressions that multiply to give

    (a) $6x + 15$ (b) $x^2 + x$ (c) $2x^2 + 5x$ (d) $2x^2 + 2x$

    (e) $10x^2$ (f) $2x^3 + 5x^2$ (g) $10x^2 + 25x$ (h) $5x^2 + 5x$

C Factorising

A $6 = 2 \times 3$

B $18 = 2 \times 3 \times 3$

C $20 = ?$

D $4x + 20 = 4(x + 5)$

E $12y - 6 = ?$

F $18k + 12 = ?$

G $a^2 - 4a = ?$

H $4x + 2x^2 = ?$

I $h^3 - h = ?$

J $6p - 9p^2 = ?$

C1 Factorise

(a) $3m + 12$

(b) $4n - 6$

(c) $15 - 10p$

(d) $8q - 4$

(e) $9a^2 - 6$

(f) $b^2 + 3b$

(g) $5c + c^2$

(h) $d^2 - 7d$

(i) $11x - x^2$

(j) $y^2 + y$

(k) $w^3 - 2w$

(l) $7h^3 + h$

C2 Find the missing expressions in these statements

(a) $2k(\blacksquare) = 2k^2 + 6k$

(b) $3p(\blacksquare) = 3p^2 - 6p$

(c) $5n(\blacksquare) = 20n^2 + 5n$

(d) $2x(\blacksquare) = 6x^2 + 4x$

(e) $h^2(\blacksquare) = h^3 + 2h^2$

(f) $y^2(\blacksquare) = 2y^3 + 3y^2$

C3 Factorise completely

(a) $3b^2 + 6b$

(b) $10a + 2a^2$

(c) $5d^2 - 15d$

(d) $21c + 7c^2$

(e) $6k^2 + 8k$

(f) $9h + 6h^2$

(g) $9x^2 - 3x$

(h) $2y + 10y^2$

C4

E	L	R	G	U	A	S	B	W	D	H	O
5	7	n	$2n$	$3n$	$n + 1$	$n - 1$	$2n + 3$	$2n - 3$	$3n + 2$	$n^2 + 2$	$n^2 - 2$

Fully factorise each expression below as the product of two factors.
Use the code above to find a letter for each factor.

Rearrange the letters in each part to spell an animal.

(a) $7n + 7$ $2n^2 - 3n$ $3n^2 - 3n$

(b) $14n - 21$ $5n - 5$ $5n + 5$

(c) $10n + 15$ $2n^2 + 2n$ $3n^2 + 2n$

(d) $5n^2 + 10$ $15n + 10$ $2n^3 + 4n$ $2n^3 - 4n$

C5 Factorise completely

(a) $n^3 + 4n^2$

(b) $6n^5 - n^3$

(c) $5n^3 + 7n^2$

(d) $2n^3 + 2n^2$

*C6 (a) Factorise $3n + 6$.

(b) Explain how the factorisation tells you that $3n + 6$ will be a multiple of 3 for any integer n.

When factorising an expression, remember to factorise completely.

Examples

$3a - 12 = 3(a - 4)$

$8k + 2k^2 = 2(4k + k^2)$
$= 2k(4 + k)$

This is unfinished because $4k + k^2 = k(4 + k)$

$6x - 15x^2 = 3x(2 - 5x)$

D *More than one letter*

Examples of substitution

Find the value of $3a^2b + b$ when $a = 4$ and $b = 5$

$3a^2b + b = 3 \times a^2 \times b + b$
$ = 3 \times 4^2 \times 5 + 5$
$ = 3 \times 16 \times 5 + 5$
$ = 240 + 5$
$ = 245$

Find the value of $5xy^2 - 2x$ when $x = 3$ and $y = 2$

$5xy^2 - 2x = 5 \times x \times y^2 - 2 \times x$
$ = 5 \times 3 \times 2^2 - 2 \times 3$
$ = 5 \times 3 \times 4 - 2 \times 3$
$ = 60 - 6$
$ = 54$

D1 Find the value of each expression when $a = 2$ and $b = 5$.

(a) $2ab + a$
(b) $5a^2 - b$
(c) $a^2 + b^2 - 1$
(d) $3a + 2b^2 - 5b$
(e) $3ab + b^2 - a^2$
(f) $(ab)^2$
(g) a^2b
(h) ab^2
(i) $a^3 + b$

D2 Find the value of each expression when $x = 3$, $y = 4$ and $z = 6$.

(a) $x + y + z$
(b) $xy + yz$
(c) $3xy - z^2$
(d) $x^2 + 2x - y$
(e) $xy^2 - 3y + z$
(f) $\frac{xy}{z}$

D3 Solve the puzzles on sheet P56.

Examples of collecting like terms

$3x + 2y + 7x - 5y = 3x + 7x + 2y - 5y$
$ = 10x - 3y$

$ab + 4a^2 + 4ab - b - 7a^2 = ab + 4ab + 4a^2 - 7a^2 - b$
$ = 5ab - 3a^2 - b$

Neither of these can be simplified any further.

D4 Simplify the following expressions by collecting like terms.

(a) $2a + b + 3a - 5b$
(b) $ab + b^2 + 5ab + 3b^2 + 2$
(c) $a^2 + 5a + 2 - 8a$
(d) $8b - a^2 + 2b + 3a^2 - 7$

D5

$\boxed{A \quad x^2 - x}$ $\boxed{B \quad 3x - y}$ $\boxed{C \quad 5x + y}$ $\boxed{D \quad y^2 + x^2}$ $\boxed{E \quad 5y - y^2}$

Find pairs of the above expressions that add to give

(a) $y^2 + 2x^2 - x$ (b) $5y + x^2$ (c) $x^2 + 2x - y$

(d) $x^2 + 4x + y$ (e) $8x$ (f) $5x + 6y - y^2$

Examples of multiplying

$3a \times 4b = 3 \times a \times 4 \times b$ $\qquad = 3 \times 4 \times a \times b$ $\qquad = 12ab$

$5zy \times 3z^2 = 5 \times z \times y \times 3 \times z \times z$
$\qquad\quad = 5 \times 3 \times z \times z \times z \times y$
$\qquad\quad = 15z^3y$

D6 Find the result of each multiplication in its simplest form.

(a) $b \times 2a$ (b) $2x \times 3y$ (c) $4p \times 3q$

(d) $5c \times d$ (e) $5m \times 4n$ (f) $3v \times 5w$

D7 Find the result of each multiplication in its simplest form.

(a) $2a^2 \times 3b$ (b) $3xy \times 4x$ (c) $2pq \times 3pq$

(d) $4cd \times 5c^2$ (e) $2m^2n^2 \times 7mn^2$ (f) $4vw^4 \times 6v^3w^2$

D8 Find the missing expression in each statement.

(a) $5p \times \blacksquare = 10pq$ (b) $\blacksquare \times 7n = 21mn$ (c) $3a^2 \times \blacksquare = 9a^2b$

(d) $\blacksquare \times 2y^2 = 10x^2y^2$ (e) $3cd^2 \times \blacksquare = 15c^3d^2$ (f) $\blacksquare \times 7vw^3 = 28v^3w^4$

D9

$\boxed{2a}$ $\boxed{3b}$ $\boxed{5ab}$ $\boxed{3a^2}$ $\boxed{2b^2}$ $\boxed{5a^2b}$ $\boxed{ab^3}$

Find pairs of the above expressions that multiply to give

(a) $6ab$ (b) $10ab^3$ (c) $15a^4b$

(d) $9a^2b$ (e) $2ab^5$ (f) $25a^3b^2$

D10 Solve the 'cover-up' puzzle on sheet P57.

Powers

$(2pq)^3 = 2pq \times 2pq \times 2pq$
$\qquad\quad = 2 \times p \times q \times 2 \times p \times q \times 2 \times p \times q$
$\qquad\quad = 2 \times 2 \times 2 \times p \times p \times p \times q \times q \times q$
$\qquad\quad = 8p^3q^3$

D11 Expand and simplify

(a) $(3pq)^2$ (b) $(2vw)^3$ (c) $(5x^2y)^2$ (d) $(2a^2b^3)^4$

Examples of division

$$\frac{8ab}{2b} = \frac{8 \times a \times b}{2 \times b}$$

$$= \frac{\overset{4}{\cancel{8}} \times a \times \cancel{b}}{\cancel{2} \times \cancel{b}}$$

$$= 4a$$

$$\frac{12a^2b}{4ab} = \frac{12 \times a \times a \times b}{4 \times a \times b}$$

$$= \frac{\overset{3}{\cancel{12}} \times a \times \cancel{a} \times \cancel{b}}{\cancel{4} \times \cancel{a} \times \cancel{b}}$$

$$= 3a$$

$$\frac{a^5b^3}{a^2b^4c} = \frac{a \times a \times a \times a \times a \times b \times b \times b}{a \times a \times b \times b \times b \times b \times c}$$

$$= \frac{a \times a \times a \times \cancel{a} \times \cancel{a} \times \cancel{b} \times \cancel{b} \times \cancel{b}}{\cancel{a} \times \cancel{a} \times b \times \cancel{b} \times \cancel{b} \times \cancel{b} \times c}$$

$$= \frac{a \times a \times a}{b \times c}$$

$$= \frac{a^3}{bc}$$

D12 Simplify

(a) $\dfrac{10pq}{2p}$

(b) $\dfrac{12xy}{6y}$

(c) $\dfrac{18mn}{6m}$

(d) $\dfrac{8ab^2}{2a}$

(e) $\dfrac{6p^2q}{2p}$

(f) $\dfrac{8x^2y}{4xy}$

(g) $\dfrac{12m^2n}{3mn}$

(h) $\dfrac{16a^3b^2}{4a^2b^2}$

D13

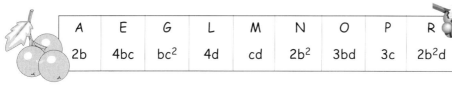

A	E	G	L	M	N	O	P	R
$2b$	$4bc$	bc^2	$4d$	cd	$2b^2$	$3bd$	$3c$	$2b^2d$

Simplify each expression below as far as you can.
Use the code above to find a letter for each expression.

Rearrange each set of letters to spell a fruit.

(a) $\dfrac{8cd}{2c}$ $\quad \dfrac{12bc}{3} \quad$ $\dfrac{15cd}{5d}$ $\quad \dfrac{4bc}{2c} \quad$ $\dfrac{9c^2b}{3cb}$

(b) $\dfrac{18b^2d}{6b}$ $\quad \dfrac{10ab^2}{5a} \quad$ $\dfrac{20b^2c^2}{5bc}$ $\quad \dfrac{12cd^5}{3cd^4} \quad$ $\dfrac{5c^2d}{5c}$

(c) $\dfrac{21bc^2}{7bc}$ $\quad \dfrac{32b^2c^4}{8bc^3} \quad$ $\dfrac{14bcd}{7cd}$ $\quad \dfrac{6b^5d}{3b^3} \quad$ $\dfrac{7b^5c^3}{7b^4c}$

(d) $\dfrac{5b^6c^7}{5b^5c^5}$ $\quad \dfrac{16b^5d^2}{8b^3d} \quad$ $\dfrac{8b^6c^3}{4b^5c^3}$ $\quad \dfrac{2db^5}{b^3d} \quad$ $\dfrac{6b^2d^7}{2bd^6}$ $\quad \dfrac{20b^2c^2d}{5bcd}$

D14 Simplify

(a) $\dfrac{ab}{bc}$

(b) $\dfrac{ab}{6b}$

(c) $\dfrac{6xy}{y^2}$

(d) $\dfrac{16mn^2}{8mn^3}$

(e) $\dfrac{p^6q^4}{p^3q^6}$

(f) $\dfrac{10ab^2c}{15abc}$

(g) $\dfrac{3a}{9a^2b}$

(h) $\dfrac{4gh^2}{8g^2h^2}$

D15 Simplify

(a) $\dfrac{2hk \times 6hk^2}{3h}$

(b) $\dfrac{4m^2n \times 5mn^2}{10m^3}$

(c) $\dfrac{4x^3y \times 3xz^3}{6x^2y}$

E Expanding and factorising

Examples

Expand $3a(b + a)$

$$3a(b + a) = 3a \times b + 3a \times a$$
$$= 3ab + 3a^2$$

Factorise completely $6a^2 - 9ab$

$$6a^2 - 9ab = a(6a - 9b)$$
$$= 3a(2a - 3b)$$

This is unfinished because $6a - 9b = 3(2a - 3b)$

E1 Expand

(a) $4(a + b)$ (b) $3(x - y)$ (c) $5(m + 2n)$ (d) $4(3a - 5b)$

(e) $h(k + 1)$ (f) $p(p - q)$ (g) $b(3a + 5)$ (h) $c(2d + 3c)$

E2 Factorise

(a) $3p + 3q$ (b) $5k - 5h$ (c) $2a + 6b$ (d) $6n - 9m$

(e) $ab + a$ (f) $x^2 - xy$ (g) $7pq + 9p$ (h) $4ab + 5a^2$

E3 Expand

(a) $2a(b - a)$ (b) $3x(2y + 5)$ (c) $5m(n + 2m)$ (d) $xy(y - 1)$

(e) $ab(5a + 3b)$ (f) $p^2(3p - 1)$ (g) $3xy(2x + 9)$ (h) $2y^2(5x - 1)$

E4 Factorise completely

(a) $3m^2 - 3mn$ (b) $2x^2 + 4xy$ (c) $3p^2 - 6pq$ (d) $10mn + 15m^2$

(e) $ab^2 + 3ab$ (f) $x^2y - xy$ (g) $6y^2z + 10y^2$ (h) $10k^2h - 5k^2$

(i) $2x^2y + 2xy^2$ (j) $3a^2b - 15ab$ (k) $8hk + 4hk^2$ (l) $10p^2q + 5pq^2$

E5

E	H	P	S	O	A	I	L	G	R	T	U	N
5	$2a$	$3a$	$2b$	$7b$	a^2	ab	$3b^2$	$a + b$	$a - 5b$	$2a - b$	$ab + 1$	$2a + 3b$

Fully factorise each expression below as the product of two factors.
Use the code above to find a letter for each factor.

Rearrange each set of letters to spell a bird.

(a) $3a^2 - 15ab$ $2a^3 - a^2b$ $7ab - 35b^2$

(b) $4a^2 - 2ab$ $2a^2b + 2a$ $2ab - 10b^2$

(c) $7ab + 7b^2$ $5a - 25b$ $2ab^2 + 2b$

(d) $4ab - 2b^2$ $3b^2a + 3b^3$ $a^3 - 5a^2b$ $2a^2b + 3ab^2$

E6 Factorise completely

(a) $a^3b + 2a^2b^2$ (b) $7m^2n^2 + mn^2$ (c) $6x^2y^2 - 2xy^3$ (d) $3p^4q + 12p^2q^2$

F Formulas

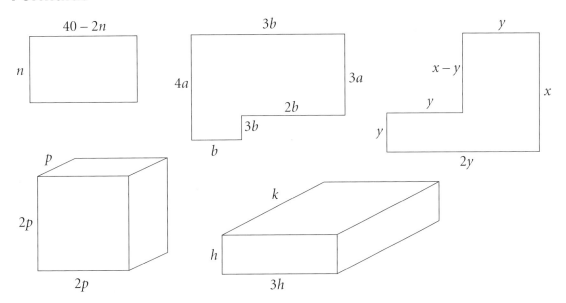

F1 (a) Find a formula for the perimeter of each shape below.
Use P to stand for the perimeter each time. (Each formula begins $P = \ldots$.)

(b) Find a formula for the area of each shape.
Use A to stand for the area each time.

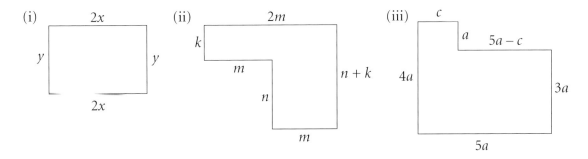

(i) [rectangle: $2x$ top, y left, y right, $2x$ bottom]

(ii) [L-shape: $2m$, k, m, n, $n+k$, m]

(iii) [shape: c, a, $5a - c$, $4a$, $3a$, $5a$]

F2 Find a formula for the volume of each prism.
Use V to stand for the volume each time.

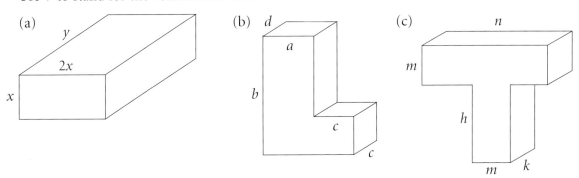

(a) [prism: y, $2x$, x]

(b) [L-prism: d, a, b, c, c]

(c) [T-prism: n, m, h, m, k]

*F3 (a) Find a formula for the volume of each prism.
Use V to stand for the volume each time.

(b) Find a formula for the surface area of each prism.
Use S to stand for the surface area each time.

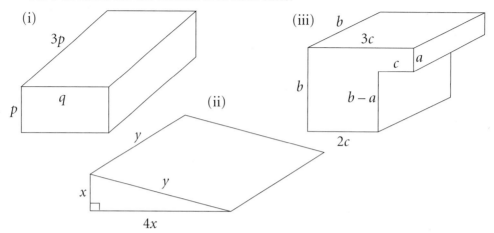

(i)

(ii)

(iii)

Test yourself with these questions

T1 Write an expression, as simply as possible, for
the perimeter of this shape.

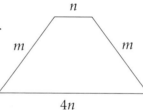

T2 Simplify

    (a) $4x + 2y + 3x - 6y$ (b) $a^2 - 4a - 1 + 3a - 2$

    (c) $3h \times 5k$ (d) $3a^2b \times 2ab$

T3 Multiply out

    (a) $2(p + q)$ (b) $x(x - 2)$ (c) $p^2(3p - p^4)$ (d) $5h(2h - k)$

T4 Factorise completely

    (a) $2a + 6b$ (b) $8x - 2xy$ (c) $x^2 + 6x$

    (d) $3p^2 + 9pq$ (e) $4n^2 - 8n$ (f) $7a^2b + 21ab^4$

T5 (a) Expand $x(5x^2 + 1)$ (b) Simplify $2a^3y^2 \times 5ay^4$

    (c) Factorise completely $6x^2y - 9xy^2$

T6 Simplify $\dfrac{a^6c^4}{a^2c^5}$ AQA(NEAB)1998

T7 Simplify $\dfrac{3x^5y^3}{6xy^2}$

30 Triangles and polygons

You will revise

◆ the names and properties of special types of triangles and quadrilaterals

The work will help you

◆ understand and use angle properties of triangles, quadrilaterals and other polygons

A Describing shapes

A1 This regular hexagon has been split into two trapeziums.

Draw sketches to show how a regular hexagon
can be split into each of the following.
Use triangular dotty paper if you like.

(a) Three rhombuses

(b) Six equilateral triangles

(c) Four trapeziums

(d) A kite and two isosceles triangles

(e) A rectangle and two isosceles triangles

(f) An equilateral triangle and three isosceles triangles

(g) An isosceles triangle and two trapeziums

A2 This square has been divided into
two right-angled triangles and a parallelogram.

Draw sketches to show how a square
can be split into each of the following.
Use square dotty paper if you like.

(a) Two right-angled triangles

(b) Three right-angled triangles

(c) Two isosceles triangles and two trapeziums

(d) A kite and two right-angled triangles

(e) An isosceles triangle and two right-angled triangles

(f) Two trapeziums

(g) A square and four right-angled triangles

A3 A quadrilateral has rotation symmetry of order 4.

(a) What is the name of the quadrilateral?

(b) Describe some other special properties that it has.

B Angle properties – a reminder

Angles on a line add up to 180°. $a + b = 180°$	Angles round a point add up to 360°. $c + d + e = 360°$	Vertically opposite angles are equal. $f = g$ and $h = i$ 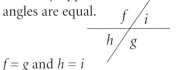
Corresponding angles (made with parallel lines) are equal. $j = k$	Alternate angles (made with parallel lines) are equal. $l = m$	Angles between parallel lines, on the same side of a line crossing them, add up to 180°. $n + o = 180°$

B1 Find the angles marked with letters.

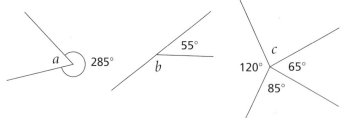

B2 Find the angles marked with letters.

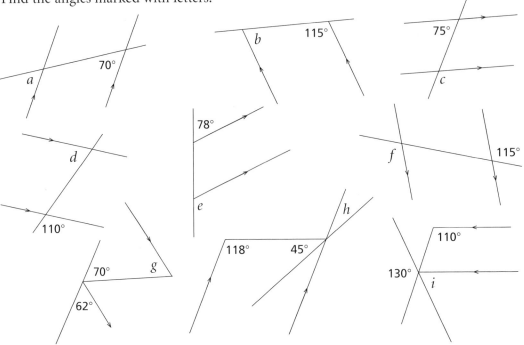

C Angles of a triangle

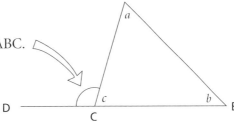

Angle ACD is an **exterior angle** of triangle ABC.

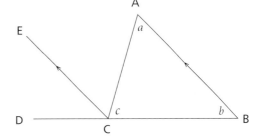

The line CE has been drawn parallel to AB.

- What is the value of angle ECD ? Why?
- What is the value of angle ACE ? Why?
- What does the following statement mean and why is it true?

    An exterior angle of a triangle is equal to the sum of the other two interior angles.

- How does this show that **the angles of a triangle add up to 180°** ?

C1 Work out the missing angles.

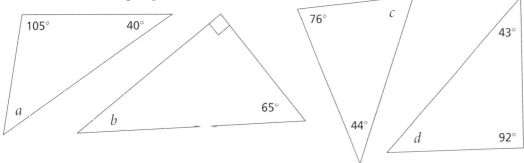

C2 Work out the missing angles in these isosceles triangles.

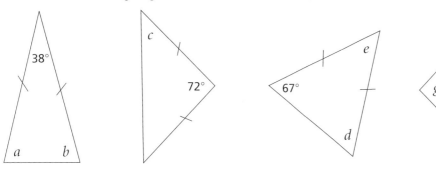

C3 Work out the angles marked with small letters.
Explain how you worked out each angle.

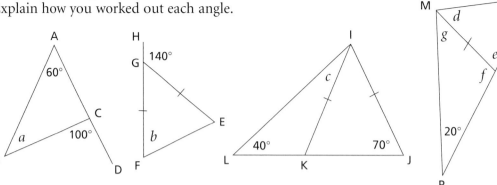

Ⓓ *Polygons*

D1 You can split a quadrilateral into two triangles like this.

(a) What is the sum of the 'black' angles?

(b) What is the sum of the 'white' angles?

(c) What is the sum of the interior angles of a quadrilateral?

D2 Find the angles marked with letters.
Explain how you worked out each angle.

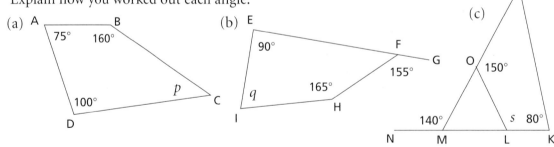

D3 (a) Draw any polygon and record how many sides it has.

Choose a vertex. Draw straight lines from it to all the other vertices.

(b) How many triangles have you made?

(c) What is the total of all the angles in the triangles?

(d) From your answer to (c) write the total of all the interior angles of your polygon.

(e) Do (a) to (d) again, for a polygon with a different number of sides.

(f) Suppose the polygon you start with has n sides.
Go through (a) to (d) again to find a formula for the total of the interior angles.

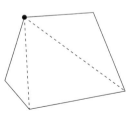

The sum of the interior angles of a polygon with n sides is $180(n - 2)°$.

D4 What is the sum of the interior angles of an 11-sided polygon?

D5 For each of these polygons,
    (i) record the number of sides
    (ii) work out the sum of the interior angles
    (iii) work out the missing angle

(a)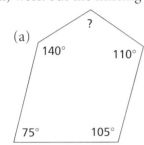

(b)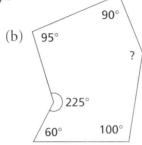

(c)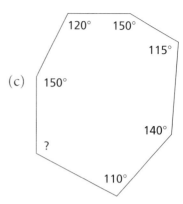

D6 Each of these polygons has one line of reflection symmetry.
Work out the missing angles.

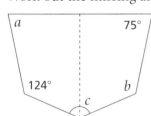

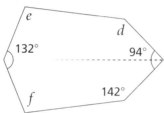

A **regular** polygon has all its sides equal and all its angles equal.

D7 Find the size of each interior angle of
    (a) a regular hexagon
    (b) a regular nonagon (nine sides)

As with a triangle, if you extend a side of a polygon,
the angle made is called an **exterior angle**.

D8 If a pencil is moved around the sides of a polygon,
at each vertex it turns through the exterior angle.

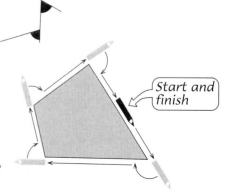

    (a) When the pencil gets back to where it started from,
it will be pointing in the same direction as before.
What angle has it turned through?

    (b) What is the sum of the exterior angles of a polygon?

The sum of the exterior angles of a polygon is 360°.
This is true however many sides it has.

D9 For each of these polygons,

(i)   Find the missing exterior angle or angles.

(ii) Work out the interior angle at each vertex.

(iii) Work out the total of the interior angles.
Use the formula from question D8 to check whether
this total agrees with the number of sides of the polygon.

(a)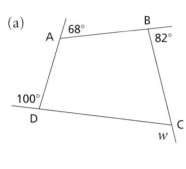

(b)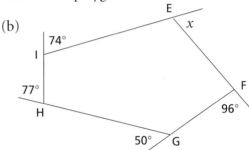

(c)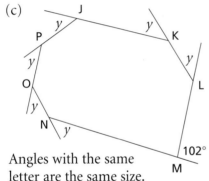

Angles with the same
letter are the same size.

(d)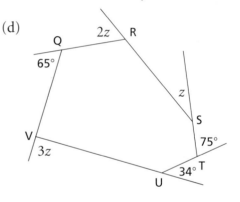

D10 This is a regular 10-sided polygon (a decagon).
Each angle is marked *e*.

Calculate the size of *e*.

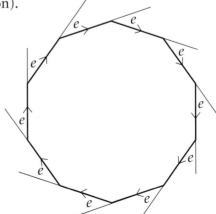

D11 Calculate the size of each exterior angle of a regular polygon with

    (a) 18 sides   (b) 24 sides   (c) 30 sides   (d) 45 sides

D12 This is a close-up a vertex of the decagon in D10.

    What is the size of the interior angle, i?

D13 Calculate the interior angle for each regular polygon in D11.

D14 Each exterior angle of a certain regular polygon is 30°.
How many sides must this polygon have?

D15 How many sides does a regular polygon have if each exterior angle is

    (a) 9°   (b) 24°   (c) 10°   (d) 18°

D16 How many sides does a regular polygon have if each interior angle is

    (a) 135°   (b) 108°   (c) 175°   (d) 174°

D17 A regular polygon has n sides. Write an expression in n for

    (a) the size of an exterior angle.   (b) the size of an interior angle.

E *Mixed questions*

When you do questions of this kind, it is a good idea to sketch the diagram
and mark the values of angles on it as you work them out.

E1 GHIJ is a rectangle.
JILK is a kite.
GI and JL are parallel.
HL is a straight line.

    (a) Find the size of angle KJL.

    (b) Find the size of angle KLI.

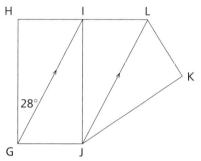

E2 This is a regular nonagon.

    (a) Triangle ADG is equilateral.
         Calculate angle x.

    (b) Calculate angles y and z.

    (c) Explain how you know that AG
         is parallel to IH.

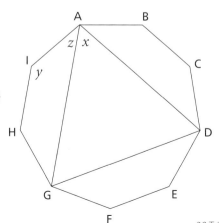

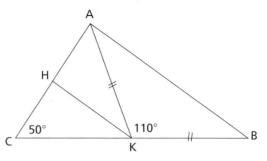

*E3 (a) Calculate angle CAK.

(b) What value must angle AHK have for HK to be parallel to AB?

Test yourself with these questions

T1 Calculate the lettered angles.

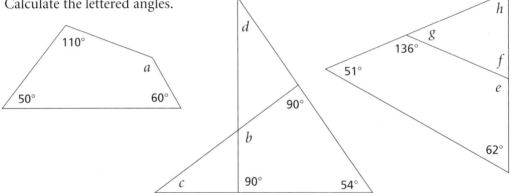

T2 Sketch these diagrams.
Fill in the sizes of all the angles.

(a)

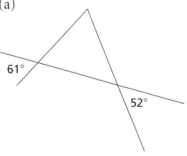

(b)

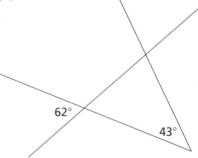

T3 Find the missing angle.

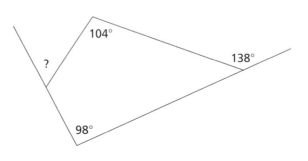

You should know
◆ how to plot points and draw the graph of a straight line, given its equation
◆ how to rearrange simple formulas

You will learn how to
◆ find the equation of any straight line
◆ decide when two straight lines are parallel

A Straight lines and equations

A1 (a) Copy and complete this table.

(b) On the same set of axes, plot the graphs of $y = x$, $y = 2x$, $y = 3x$, and $y = ^-4x$ for values of x between $^-1$ and 3.

(c) What is the gradient of each line?

(d) What do you notice? Can you explain it?

x	$^-1$	0	1	2	3
$2x$					
$3x$					
^-4x					

A2 (a) Copy and complete this table.

(b) On the same set of axes, plot the graphs of $y = x + 1$, $y = 4 + x$ and $y = x - 2$ for values of x between $^-1$ and 3.

(c) What is the gradient of each line?

(d) Where does each line cross the y-axis?

(e) What do you notice? Can you explain it?

x	$^-1$	0	1	2	3
$x + 1$					
$4 + x$					
$x - 2$					

A3 Use sheet P58 for this question.

(a) For each set of equations on the right,

   (i) draw the graphs on the same set of axes

   (ii) find the gradient of each line

   (iii) write down where each line crosses the y-axis

(b) Briefly summarise your results.

A
$y = x + 3$
$y = 3 + 2x$
$y = 3x + 3$

B
$y = 2x + 5$
$y = 2x - 1$
$y = 2x + 2$

C
$y = ^-x + 1$
$y = ^-x + 5$
$y = ^-x - 3$

D
$y = ^-2x + 1$
$y = ^-2x + 2$
$y = ^-2x - 1$

A4 (a) Where do you think the line with equation $y = 2x + 1$ crosses the y-axis?

(b) What do you think is the gradient of the line with equation $y = 2x + 1$?

ⓑ *Finding equations*

The equation of a straight line can be written in the form $y = mx + c$.
m is the gradient and c is the y-intercept.

The y-intercept is wh
the line cuts the y-ax

Examples

A

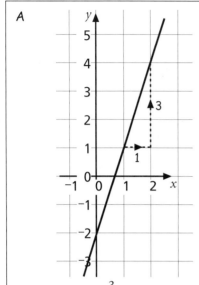

The gradient is $\frac{3}{1} = 3$.
The y-intercept is $^-2$.
So the equation of the line is $y = 3x - 2$.

B

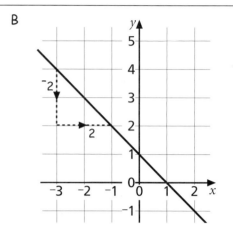

The gradient is $\frac{^-2}{2} = {}^-1$.
The y-intercept is 1.
So the equation of the line is $y = {}^-1x + 1$.
We usually write this as $y = {}^-x + 1$
or $y = 1 - x$.

B1 For each line below   (a) Find the gradient and y-intercept.

(b) Write down its equation.

A

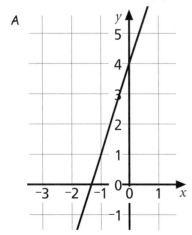

B

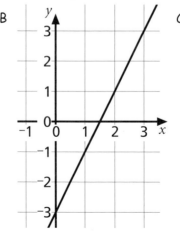

C

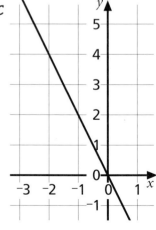

B2 (a) Plot the points $(^-2, {}^-9)$ and $(3, 11)$ on a set of axes.

(b) Join the points and find the equation of this line.

B3 For each line,

    (a) find the gradient and write it as a decimal

    (b) find its equation

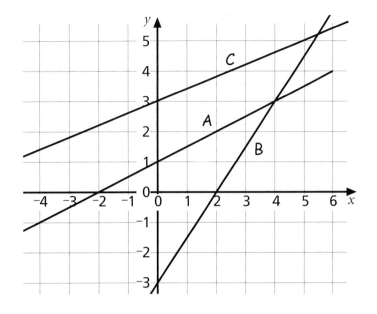

B4 (a) Plot the points (⁻5, ⁻6) and (5, 2) on a set of axes.

    (b) Join the points and find the equation of this line.

B5 Find the equation of each line below.

(a)
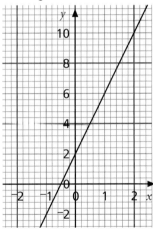

(b)
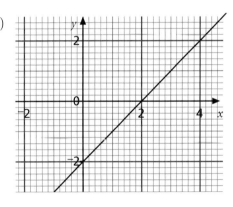

(c)
(d)
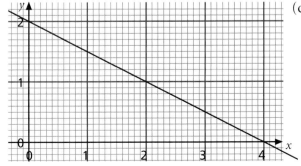
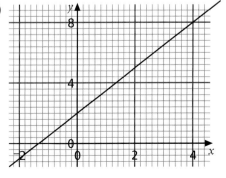

B6 Write down the equation of the line with gradient 8 that crosses the y-axis at $(0, 10)$.

B7 (a) What is the gradient of the line with equation $y = 4x - 9$?

(b) Where does it cross the y-axis?

B8 This diagram shows three parallel lines.

(a) Write down equations for lines A and B.

(b) Write down an equation of any other line that is parallel to these three.

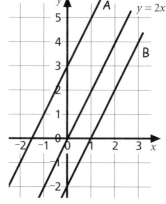

B9 What is the equation of the line parallel to $y = 3x + 2$ that crosses the y-axis at $(0, -1)$?

B10 The lines labelled A to C match these equations.

$y = 2x + 5$

$y = x + 5$

$y = x - 2$

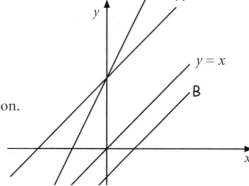

Match each line to its correct equation.

B11 This question is on sheet P59.

*__B12__ The lines labelled P to T match these equations.

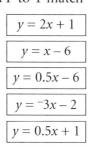

$y = 2x + 1$

$y = x - 6$

$y = 0.5x - 6$

$y = {}^-3x - 2$

$y = 0.5x + 1$

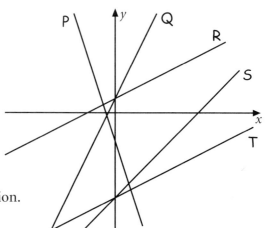

Match each line to its correct equation.

ℂ *Including fractions*

Sometimes the gradient of a line is a fraction that cannot be written as a simple decimal.

Example

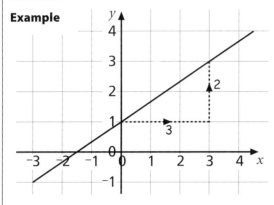

The gradient of this line is $\frac{2}{3}$.

The y-intercept is 1.

So the equation can be written

$y = \frac{2}{3}x + 1$ or $y = \frac{2x}{3} + 1$.

As a decimal $\frac{2}{3} = 0.6666 \dots$ or $0.\dot{6}$

We don't usually write recurring decimals in equations.

C1 For each line on the right

   (a) Find the gradient as a fraction.

   (b) Find the y-intercept.

   (c) Write down its equation.

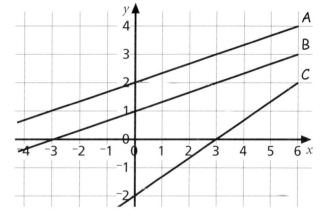

C2 Which of these lines is steeper: $y = \frac{3}{4}x - 2$ or $y = \frac{1}{4}x + 5$?

C3 What is the equation of the line joining $(0, 4)$ to $(3, 0)$?

C4 (a) Match the equations below to give four pairs of parallel lines.

   A $\boxed{y = \frac{1}{2}x + 1}$ B $\boxed{y = \frac{1}{3}x + 5}$ C $\boxed{y = -\frac{3}{4}x + 3}$ D $\boxed{y = \frac{x}{5} - 1}$ E $\boxed{y = 9 + \frac{1}{2}x}$

      F $\boxed{y = \frac{x}{3} - 2}$ G $\boxed{y = \frac{1}{5}x + 3}$ H $\boxed{y = \frac{4}{3}x + 2}$ I $\boxed{y = 1 - \frac{3}{4}x}$

   (b) Which equation is the odd one out?

C5 What is the gradient of the line with equation $y = \frac{x}{6} + 5$?

Ⅾ *Rearranging*

A $y = x - 4$ B $y - 2x = 3$ C $y = 7 - x$ D $y = 2x + 1$

E $y = 1 - 2x$ F $y - x = 3$ G $y + 2x = 5$

- Which pairs of these equations give parallel lines?

D1 Find the gradient of each of these lines.

(a) $y - x = 5$ (b) $y - 2x = 4$ (c) $3x + y = 6$ (d) $2x = y + 3$

D2 Match the equations below to give four pairs of parallel lines.

A $y - x = 2$ B $y - 3x = 8$ C $y = 5 - 3x$ D $y + 3x = 1$

E $y = 4x - 2$ F $y = x - 6$ G $y - 4x = 5$ H $y = 5 + 3x$

Sometimes you must **divide** the equation of a straight line to give it in the form $y = mx + c$.

Examples

A line has equation $2y = 4x - 8$.

$$2y = 4x - 8 \quad (\div 2)$$
$$y = 2x - 4$$

So the gradient is 2 and the y-intercept is ⁻4.

A line has equation $3y + 15x = 3$.

$$3y + 15x = 3 \quad (\div 3)$$
$$y + 5x = 1 \quad \text{(rearrange)}$$
$$y = 1 - 5x \quad \text{(rearrange)}$$
$$y = {}^-5x + 1$$

So the gradient is ⁻5 and the y-intercept is 1.

D3 Find the gradient of each of these lines.

(a) $2y = 4x + 10$ (b) $3y = 9 - 12x$ (c) $2y - 8x = 12$ (d) $4y + 12x = 8$

D4 Match the equations below to give four pairs of parallel lines.

A $3y = 3x + 21$ B $4y = 8x - 4$ C $2y - 6x = 10$ D $y = {}^-x + 3$

E $y = 2x - 3$ F $y = x + 5$ G $5y + 5x = 10$ H $y = 1 + 3x$

D5 Find the gradients and y-intercepts of each of these lines.

(a) $2y = x + 6$ (b) $3y = x$ (c) $4y = x + 8$

D6 Which two of these lines are parallel to $y = \frac{1}{2}x - 3$?

A $2y = x + 2$ B $2x = y + 10$ C $2y + 8 = x$ D $y - 2x = 9$

*D7 Show that $5x = y - 1$ and $6y - 30x = 6$ are equations for the same line.

E *Lines of best fit*

A pan of water was heated and the temperature measured at various intervals.
The results were plotted and a line of best fit drawn.
How can we find the equation of the line of best fit?

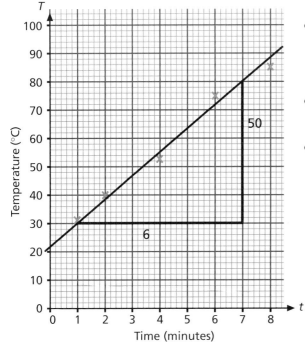

- Choose two convenient points on the line and use them to find the gradient:

$$\frac{50}{6} = 8.3 \text{ (to 1 d.p.)}$$

- Find where it cuts the vertical axis:

    The line cuts this axis at about 22

- So the equation of the line of best fit is approximately:

$$T = 8.3t + 22$$

where T is the temperature in °C and t is the time in minutes.

E1 Nigel lit a small candle.
He put a beaker upside down over the candle and counted how many seconds it took for the flame to go out.

He repeated the experiment with larger containers and plotted his results on a graph (on sheet P60).

(a) Draw the line of best fit on the graph.

(b) Choose two points **on your line** and use them to find the gradient, correct to 2 d.p.

(c) Find where the line of best fit cuts the vertical axis.

(d) Hence find an approximate equation for the line of best fit.

(e) Use your equation to estimate how long it would take for the flame to go out in a beaker with volume 1200 ml.

E2 Some more sets of points are shown on sheet P61 and P62.

(a) Draw a line of best fit on each graph.

(b) Estimate the gradient and vertical intercept of each line.

(c) Find an approximate equation for each line.

(d) Answer the questions beneath each graph.

Test yourself with these questions

T1 (a) Find the gradient of this line.

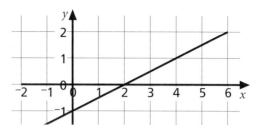

(b) Write down the equation of the line.

T2 (a) Plot the points $(0, 5)$ and $(4, 13)$ on a set of axes.

(b) Find the equation of the line through these points.

T3 A is the point $(0, 3)$ and B is the point $(3, 9)$.

(a) Calculate the gradient of the line AB.

(b) Write down the equation of the line AB.

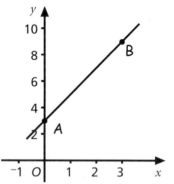

AQA(NEAB)1997

T4 A graph of the equation $y = ax + b$ is shown. Find the values of a and b.

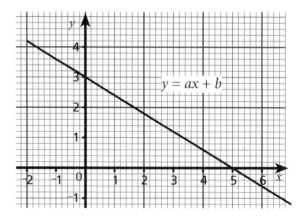

AQA(SEG)1998

T5 Here are the equations of 5 straight lines. They are labelled from A to E.

Which two lines are parallel?

A	$y = 2x + 1$
B	$y = 1 - 2x$
C	$2y = x - 1$
D	$2x - y = 1$
E	$x + 2y = 1$

Edexcel

Review 6

1 Using a ruler and compasses only, construct an angle of 105°.

2 For each of these, write down the value of n.
   (a) $4^n = 16$ (b) $9^n = 9$ (c) $n^4 = 81$ (d) $n^5 = 1$
   (e) $10^n = 10\,000$ (f) $2^n = 256$ (g) $10^n = 1\,000\,000$ (h) $n^{-2} = 0.01$

3 This diagram gives information about a restored steam railway.
   The oval boxes give heights of the track above sea level.

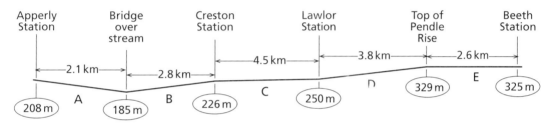

   (a) Find the gradient of each section of track, as a decimal, to 3 d.p.
   (b) Which section has the steepest gradient?

4 Find the value of each of these.
   (a) $(4^3)^2$ (b) $(3^2)^2$ (c) $(6^2)^1$ (d) $(3^5)^0$

5 Work out the missing lengths, giving your answers to 1 decimal place.

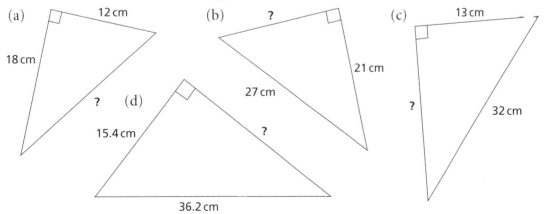

6 Simplify these expressions
   (a) $2p^3 \times 3p^2$ (b) $n^2 \times 9n$ (c) $\dfrac{q^4}{q^3}$ (d) $\dfrac{w^4 \times w^5}{w^7}$
   (e) $\dfrac{k^7}{k \times k^4}$ (f) $\dfrac{18r^5}{3r^2}$ (g) $\dfrac{8b^5}{2b^7}$ (h) $\dfrac{a^7}{a^{-9}}$

7 Find the missing length.

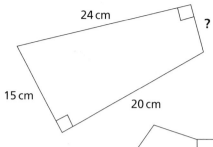

24 cm

?

15 cm

20 cm

8 This is part of a 'ring' made by alternating squares and regular pentagons.

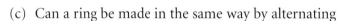

(a) Calculate the angle x.

(b) The inside of the completed ring forms a regular polygon. How many sides does it have?

(c) Can a ring be made in the same way by alternating

　(i) squares and regular hexagons

　(ii) squares and regular nonagons (nine-sided polygons)

　Describe what happens in each case.

x

9 For each of these shapes,

(i) find a formula for the perimeter ($P = \ldots$)

(ii) find a formula for the area ($A = \ldots$)

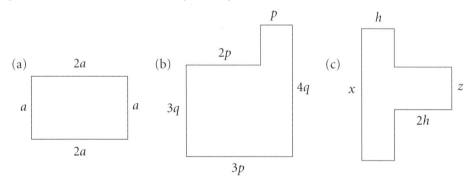

(a) 　　2a

a 　　　　a

　　2a

(b) 　　2p

　4q

3q　　

　　3p

(c) 　h

x 　　　z

　　2h

10 Side AB of the square is part of the line $y = \frac{1}{2}x + 7$.

(a) Give the equations for the other three sides of the square. (Draw the square on graph paper if this helps.)

(b) Give the equations for the diagonals of the square.

(c) Give the equation of the line going through the centre of the square, parallel to the y-axis.

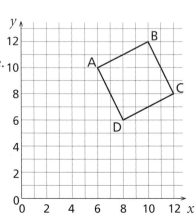

11 (a) Substitute the values $p = 4$ and $q = 1$ into the
expressions for the sides of this right-angled triangle.
Do the lengths you get obey Pythagoras's theorem?

(b) Now try $p = 3$ and $q = 2$.
Do the lengths of the sides obey Pythagoras's theorem?

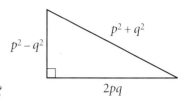

(c) What values of p and q do you need to
get a 3, 4, 5 right-angled triangle?

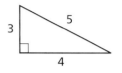

(d) Use Pythagoras to check that this triangle is right-
angled.
What values of p and q give these lengths of sides?

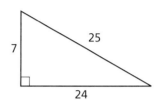

12 Use a ruler and compasses only for this question.

Draw a **large** circle in the centre of a sheet of paper.

Mark four points ABCD anywhere on the circle and
join them to produce a quadrilateral.

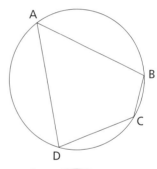

Construct the perpendicular bisector of side AB and
label point P, the mid point of AB.

Similarly, by construction, mark

  • Q, the midpoint of BC
  • R, the midpoint of CD
  • S, the midpoint of DA

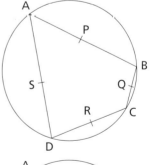

Construct a perpendicular from P to the side CD

Similarly, construct

  • a perpendicular from Q to DA
  • a perpendicular from R to AB
  • a perpendicular from S to BC

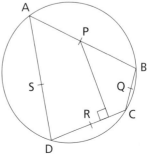